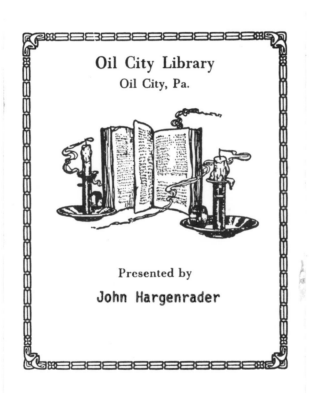

NASA SP-2005-4537

Robert C. Seamans, Jr.

PROJECT APOLLO
The Tough Decisions

Monographs in Aerospace History
Number 37

National Aeronautics and Space Administration
Office of External Relations
History Division
Washington, DC

2005

For sale by the Superintendent of Documents, U.S. Government Printing Office
Internet: bookstore.gpo.gov   Phone: toll free (866) 512-1800;   DC area (202) 512-1800
Fax: (202) 512-2250 Mail: Stop SSOP, Washington, DC 20402-0001

ISBN 0-16-074954-9

On the cover: *A Saturn rocket figuratively reaches for the Moon.*

Library of Congress Cataloging-in-Publication Data

Seamans, Robert C.

Project Apollo: the tough decisions / Robert C. Seamans, Jr.

    p. cm.

Includes bibliographical references.

   1. Project Apollo (U.S). 2. Manned space flight 3. Space flight to the moon.
I. Title.

TL789.8.U6A581653 2005

629.45'4'0973—dc22   2005003682

ISBN 0-16-074954-9

9 780160 749544

Table of Contents

List of Figures

commander; Michael Collins, Command Module pilot; and Buzz Aldrin, Lunar Module pilot, are confined to the Mobile Quarantine Facility (MQF). (NASA Image Number S69-21365)

Acknowledgments

Gene, my bride, has been patient with me for many years, 63 to be exact. In the past, I've traveled extensively, and even when home, I've had deadlines to meet and weekend activities to attend. So last January, in retirement, I asked my daughters whether it was fair to embark on another major endeavor. I explained why I wanted again to put pencil to paper. They thought I should, but only if their mother was sympathetic. Gene was not just enthusiastic; she was supportive from then until now, even when I disappeared behind closed doors for countless hours.

My output on foolscap is illegible to most, but somehow Julie Cleary, my associate, was able to interpret and transcribe the contents. There were, of course, rewrites and additions, and then a table of contents, an index, and footnotes—this last the greatest hurdle of all. The final copy is a testament to her perseverance and skill.

I also want to recognize Marie Doherty and the members of the Media Services Department in the Draper Laboratory for their assistance with three key illustrations.

This opus could not have been published without the full support of NASA's History Division, and in particular Steve Garber, with whom I'd worked previously on my book *Aiming at Targets* (NASA SP-4106, 1996). Now, as in the past, his professionalism, his calm demeanor, and his can-do attitude made this publication become a reality.

Special thanks also go to Nadine Andreassen, Steven Dick, Giny Cheong, Annette Lin, and Mike Peacock of the History Division for all their help.

Dr. Asif Siddiqi, an expert in Russian space history, also contributed a great deal by reviewing the manuscript. Thanks also to the various peer reviewers who provided much useful feedback.

Special thanks also go to the fine professionals in the NASA Headquarters Printing and Design office. Lisa Jirousek carefully edited the manuscript, Tatiana Floyd laid it out, Jeffrey McLean and Henry Spencer handled the printing, and Steve Johnson and Gregory Treese oversaw the whole effort. My hat is off to all of these people for their expert contributions.

Foreword

Robert C. Seamans, Jr., has written a uniquely comprehensive report of the Apollo Manned Lunar Landing Program. It goes well beyond the normal reporting that we have seen of the events leading to and results achieved in that major national space program. Bob Seamans has relied on his very personal involvement, responsibility, and experience during his long tenure in the top leadership of the National Aeronautics and Space Administration (NASA), first as an Associate Administrator and then as the Agency's Deputy Administrator, from less than two years after NASA was formed until January 1968, to present a detailed timeline of the key elements of NASA's extensive analyses, decisions, activities, capabilities, and responsibilities that led to the creation of the program and its outstanding success. In fact, this manuscript presents the most detailed and specific assembly of personal and archival records to identify the comments, events, meetings, decisions, and actions taken in the initiation and conduct of the program. This detailed report demonstrates NASA's broad capabilities and, despite his unassuming presentation, also shows Bob Seamans's strong contributions. Both of those demonstrated characteristics have always been clear to all of us who worked in NASA.

The report reviews the major Mercury and then Gemini precursors for the Apollo mission program and its development and mission sequence. But, very importantly, it describes the major and often complex deliberations that encouraged inputs from the broad range of informed internal Agency individuals in order to arrive at the resulting actions taken; it recognizes differences among their various views, including even sensitivities within the leadership of the Agency, and it acknowledges NASA's relationships with the President and key executive branch personnel, as well as the very important and often complex relationships with members of Congress. The process of writing this book was searching and comprehensive. The achievement of the world's first manned lunar landings, after the earlier Mercury and Gemini programs played catch-up to match the Soviet Union's advanced position, clearly established the United States' preeminence in space. Early in the book, Bob describes an extended meeting in the White House in which the President's views and those of Mr. Webb were seriously discussed. Bob tells how, through Apollo's lunar landing, NASA clearly met both President Kennedy's goal to overcome the Soviets' leadership image and James Webb's goal to use Apollo as a major part of his program to demonstrate U.S. technological preeminence.

Apparent throughout this report is the outstanding competence and capability of the NASA organization in its Centers and Headquarters. The Agency's leadership was clearly committed to providing the budget and other requirements to achieve the clearly defined program goals. The major progress in establishing the mission flight system elements and facility infrastructure was started under NASA's first Administrator, Keith Glennan, well before the Apollo mission was defined. This report shows the major new capabilities that were required in this still-new organization to achieve this objective—operational Field Centers; entirely new facility capabilities; the technology development and equipment base; the organizational strengths, including the integrated management systems; and overall in-house competence in all of the necessary areas even while the Agency relied heavily on significant outside contractor and university capabilities to implement many of the required functions. Ultimate responsibility always remained within the NASA organization. The lunar landing was an outstanding achievement that met all its goals.

A clear requirement in achieving this success was establishing the fully integrated management structure and leadership for the various elements of the program. That task obviously received major attention from NASA's top leadership, with strong emphasis on management clearly enunciated by Administrator James Webb. His focus on management was always very clear to me, especially when he said to me, "How do I make a technical man like you understand the importance of management?" He then made me a special advisory Assistant to the Administrator while I was still serving in my technical program roles. In this new position, I analyzed the need for changes in procedures and functional alignments in Headquarters. I was then appointed the Associate Administrator for Organization and Management, combining the various Agency management functions, as Bob Seamans describes. But, well before that, with the initiation of the Apollo program, there was the need to integrate the activities of the Centers and bring strong in-house NASA people together into the newly established Office of Manned Space Flight. The need to identify a strong leader was urgent. During extensive consideration by Webb, Hugh Dryden, and Seamans of various possible candidates for that position, Bob Seamans suggested and then recruited Brainerd Holmes of RCA as that leader. When he left, George Mueller was identified by Bob Seamans and was the clear leader of Apollo through its mission achievement. As the program proceeded and as conditions changed, it is apparent throughout this report that there was a continuing emphasis on management and its changing requirements.

Clearly indicated throughout this report are the very important free and open discussions and objective analysis of perceived issues, concerns, and alternative approaches, including various mission concepts, among all of the competent technical and management members of the internal staff, even if those discussions might indicate differences of opinion regarding planned approaches. This interchange was certainly strongly encouraged and pursued by Bob Seamans. The most dramatic example of that open view and the examination of alternative approaches and suggestions was the result of Bob Seamans's actions in responding to the persistent recommendations from John Houbolt that a lunar orbit rendezvous approach was superior to the then-preferred direct lunar landing flight plans even after extensive analyses had led to that preference. Bob's willingness to consider recommendations that clashed with previously approved plans led to further examination and decision in favor of what became John's very successful lunar orbit approach for the mission. This process succeeded in spite of the repeated pessimism of President Kennedy's Science Advisor about the concept and even his pessimism about the lunar landing mission more generally.

Yes, there were tragic and painful events during this period of great progress, and these are also described in Bob's report. Certainly, the assassination of President Kennedy on 22 November 1963, only six days after he had visited the launch facilities and walked around the Saturn I launch vehicle, was devastating to the entire United States, including all of us who had been involved in fulfilling his commitment to spaceflight goals. Bob Seamans's discussion of that terrible event and of his meeting and correspondence with Jacqueline Kennedy shortly after the funeral service depicts one of the warmest, most emotional situations imaginable. That period will never be forgotten. In addition, Bob reports comprehensively on the Apollo fire during ground testing in January 1967 in which Gus Grissom, Ed White, and Roger Chaffee were killed. It was a shocking and demoralizing hit to all of us in the space program and to the nation at large. President Johnson's decision to allow NASA to investigate the accident internally led to a quick, thorough, very solid report that produced the explanation for the accident and identified ready solutions in its operations. Bob Seamans reviews that entire situation in depth, but the recollection of that terrible event is still painful.

All of this very detailed information, upbeat as well as terrible, is conveyed by Bob Seamans in his factual presentation of the sequence of major activities involved and is amplified by his personal and professional anecdotes. This is truly a unique and important record of the Apollo program's achievements and the United States' demonstrated capability and technological preeminence. I hope this capability will be advanced broadly as we move forward with innovative and beneficial aeronautics, space exploration, space science, and applications activities. This book adds substantially to our knowledge base about the Apollo program's conduct and accomplishments and provides a firm path for further progress.

As one who worked closely with Robert C. Seamans during those challenging years, even though I was not directly responsible for any Apollo activities, I must add that I benefited and learned greatly from that association. And I have especially appreciated the warm friendship that developed then and has continued since.

—Harold B. Finger, NASA Associate Administrator, Office of Organization and Management, 1967–69

Chapter 1:
INTRODUCTION

This monograph presents the history of the manned space program during the time I was the general manager, from 1 September 1960 to 5 January 1968. I've outlined chronologically and in detail the steps taken from the early Mercury days, through the operational tests conducted with Gemini, to the qualification of Apollo, all against a backdrop of Soviet missions. A chapter on NASA management during my tenure follows. Then, in the final two chapters, the U.S. manned circumlunar and lunar landing missions are compared with Soviet attempts. I've also included a few thoughts on President Bush's Vision for Space Exploration. Throughout, I have tried to describe the key technical, operational, and management milestones and how key issues in each phase of the space program were resolved.

There was a subtler area that I had to face, namely, NASA's relationship with the executive branch, Congress, and the public at large.

Appointed officials must always remember that the President won his position through a national election; his appointees must support his decisions. The only alternative is resignation. Under questioning before Congress, the President's policies, programs, and budgets must be defined and their rationale explained. However, if an appointee is asked whether an item in the President's program was requested at that budget level by an agency such as NASA, it is fair to answer in the negative, which might result in larger dollar amounts for the agency for that item. However, there isn't much slack, and it only occurs during congressional hearings. The executive branch looks askance at any suspicion of an appointee's volunteering one's own views, and my testimony at times bordered on insubordination.

The most sensitive hearing occurred before a House committee on 14 April 1961, just after Gagarin's flight. A transcript of the exchange appears later in the chapter. It took place with

Congressman David King and was about a possible lunar landing by the Union of Soviet Socialist Republics (USSR) in 1967 and our capability to compete. This matter was under consideration throughout the government at that time, and my job was to stay in the background, not get out in front. The President had to be allowed time to do his fact finding and make his policy decisions unencumbered by the testimony of junior officials. I was skirting close to the margin.

At NASA, our role was to carry out the President's agenda for a manned lunar landing within the decade. However, this agenda was questioned in a rapid-fire discourse with the President in November 1962. When asked by the President whether NASA's top priority was the lunar landing, Jim Webb (then NASA's Administrator) answered no, and when questioned further, Webb said that NASA's prime objective was preeminence in space (see chapter 3). This dichotomy of views lurked in the background throughout the decade. In the crunch, both Kennedy and Johnson were squeezing the national budgets in order to fund NASA's principal objective, the lunar landing. At its peak, the Apollo Program accounted for 32 percent of the federal research and development (R&D) budget.[1] Despite his assertion, Jim Webb actually fully supported the lunar goal and used that goal to circumvent major budget revisions by Congress. On many occasions, he would staunchly tell congressional committees that if the budget were reduced by even a small amount, the option for a lunar landing within the decade would be lost.

Success or failure was more difficult for NASA to obfuscate than for most agencies of government. Press coverage was always present at the launch pad, particularly for manned missions. In the early days, liftoff was a matter of probability, at times followed by a major explosion and the destruction of both the vehicle and the pad. Mercury-Redstone once had an electronic liftoff. The capsule and booster went through the entire 15-minute mission firing its escape rocket and executing several pyrotechnic maneuvers. In the end, the parachute dropped around the rocket's carcass while it was still upright on the pad. Photographs of the sequence were both hilarious and damaging to NASA's image and morale. That was a failure clear to behold. NASA required some manner of measuring performance and progress that didn't rely entirely on what the eye could perceive. Ultimately, the project teams agreed that success was not just the opposite of calamity, but rather the achievement of all stated objectives. The general manager became the arbiter. In the early sixties, the success level was around 55 percent for all manned and unmanned missions. By the mid-sixties, the success level rose to 80 percent.[2]

NASA often had to deal with failure. In some cases, most objectives were achieved and there was little flack within the administration, on Capitol Hill, or from the media. However, the Apollo fire in January 1967 caused a major eruption, and rightly so. The President had to decide whether to establish a presidential commission or to allow NASA to investigate itself. If the investigation was in-house, there would be suspicions of a cover-up; however, a commission takes longer to establish and get up to speed. Usually, a commission has sessions that are open to the public and the press. Presidential commissions often deliberate for over a year. President Johnson took the heat and allowed NASA to conduct its own accident review. Slightly over two months' time was required, and the findings and recommendations were precise and hard-hitting.

While the investigation was in play, the accident review board was cloistered with its major effort at Cape Canaveral. There were no press releases from the board with conjecture, which is often proven incorrect. But the President, Congress, and the media required an impartial and continuing assessment of the board's progress. My job involved periodic visits to the Cape to listen to the board's deliberation, to probe a bit, and to review the data. On the return flight to Washington, I compared notes with my assistant, Dave Williamson, and prepared a report for Mr. Webb. If acceptable to him, the report would be relayed in sequence to the

1. Frederick C. Durant III, *Between Sputnik and the Shuttle, New Perspectives on American Astronautics* (San Diego, CA: American Astronautical Society, 1981), p. 165.

2. NASA illustration, *Space Flight Record* (15 March 1966) NASA image number AD66-845.

President, Congress, and the press. My first report was printed in its entirety by the *New York Times*, but the media weren't happy campers.

In the detailed discussion of the accident in chapter 4, I note that Mr. Webb and I disagreed on how much information should be forthcoming at congressional hearings. He felt that there were reasons for secrecy, partly because of our understanding with the President, partly to protect the accident review board, and partly to avoid legal and potential lawsuits. I couldn't disagree, but I thought he was zealous in the extreme. This sensitive matter was an unsettling undercurrent when testifying before Congress. Even more troublesome were background meetings with the press; they didn't always remain off the record. On one occasion, Julian Scheer, who was in charge of NASA's public affairs, asked me to join him for a luncheon with a few well-known reporters. I knew them and agreed. I was asked why the hatch wasn't immediately opened and the astronauts saved. The answer was straightforward: the hatch opened inward, and with the pressure rise in the capsule, there was a 4-ton force holding it shut. Several days later, stories appeared in the press citing a "high-ranking NASA official." According to the press, the astronauts could be seen attempting to claw their way to safety and being unable to escape because of a bad design. NASA had attempted to provide useful background; the press had not followed the rules; and I was left to hang, turning slowly in the wind. As you can imagine, this further exacerbated my relations with Jim Webb. I realized that Jim was right about keeping things confidential. It wasn't until my experience as Administrator of the Energy Research and Development Agency that I fully appreciated Jim's leadership at NASA.

Chapter 2:
EISENHOWER'S LEGACY

NASA was nearly two years old when I became Associate Administrator and general manager. Under the leadership of Administrator T. Keith Glennan and his deputy, Hugh Dryden, much had been accomplished since the Agency's establishment in 1958. The former National Advisory Committee for Aeronautics (NACA) had been welded together with the Jet Propulsion Laboratory of the California Institute of Technology, the Army Redstone Arsenal research and development team under Wernher von Braun, and parts of the Naval Research Laboratory. A more complete discussion of NASA's Centers is included in chapter 5.

NASA programs were providing interesting and useful results with a research and development budget that had grown in three years from $300 million to nearly $1 billion. The Echo balloon could be seen overhead on clear nights, and the Television Infrared Observation Satellite (TIROS) was in orbit, providing useful information for the Weather Bureau.

Seven astronauts had been recruited and trained, and they were prepared to orbit Earth. Technicians and engineers were at Cape Canaveral preparing the Mercury capsule, the Redstone and Atlas boosters, and the launch facilities for 90-minute flights around the world. The capsule could weigh no more than 4,400 pounds with either of the two boosters, and only one, the Atlas, had the power necessary for a complete orbit.[1] Plans had been discussed at an industry conference in August

---

1. Wernher von Braun and Frederick J. Ordway, *History of Rocketry and Space Travel* (New York: Thomas Y. Crowell Company, 1975), p. 212.

for an Apollo Program to include manned circum-lunar flights.

During the fall, the final Eisenhower budget was in preparation. NASA's budget request to the Bureau of the Budget (BoB)—now the Office of Management and Budget, or OMB—was a little over $1.4 billion. This figure had been whittled to $1.109 billion by Maurice Stans, head of the BoB, and his team.[2] Keith elected to try one more time for an increase, and he took me along. He first asked for an Administrator's discretionary fund of $50 million. Maury didn't give Keith time to explain. He just said, "You've got to be kidding. What else have you in mind?" Keith then discussed the need for a $10-million line item for an experimental communication satellite, despite the fact that NASA already had the Echo balloon in orbit for communications. The balloon served as a giant 100-foot-diameter reflector in space. Maury wasn't impressed. He said that was up to the communications industry. Keith explained that industry had no means for orbiting satellites. Maury responded that NASA could include $10 million in its budget as a reimbursable item. NASA could place the communication industry's satellites in orbit on a payback basis. And that's where the discussion ended.

Labor Complications at Cape Canaveral, November–December 1960

In early November, I received a frantic call from Marshall Space Flight Center (MSFC) Director Wernher von Braun. There was a potentially serious labor problem at Cape Canaveral. Complex 37 was under construction for the Saturn I then in development at Huntsville. However, the interface between the complex and the Saturn I team had to remain flexible, so there wasn't time to send the final 5 to 10 percent of the construction out for sealed bids by contractors with unionized labor. The two unions involved were the International Brotherhood of Electrical Workers, IBEW, and the United Association of Plumbers and Journeymen. I called the presidents of both unions and asked if Wernher and I could meet with them together to discuss construction at Cape Canaveral. It was agreed, and on

a sunny mid-November day, we headed to IBEW headquarters. The reception area and boardroom would have done justice to corporate America—thick carpet, large conference table, and comfortable leather chairs. After my brief introduction, Wernher gave a careful, logical, and somewhat impassioned talk about the importance of a tight schedule for the development of large boosters in the United States. He used a few graphics to explain why government personnel were required to finish off the construction of the launch facilities, 90 to 95 percent of which would have been completed by unionized firms. They seemed to understand but said that they were a democratic organization and they would appreciate our talking to the locals in Florida. Several days later, we were in a union hall, talking to the locals. As before, I went first. Early on, Wernher said, "NASA wouldn't be able to honor its commitment to the President if . . . ." At that point, he was cut off by a local voice yelling, "What president?" Wernher replied, "President Eisenhower." The response was an emphatic "Thank God we're rid of that son of a bitch." The meeting ended with my saying that we would proceed with government employees and hope we'd have the unions' support.

We had government employees work on the construction; the union struck; and on Thanksgiving afternoon, I was being called on the carpet at Keith Glennan's apartment. Secretary of Labor James P. Mitchell had called Keith and wanted to know why NASA was trying to spoil Eisenhower's labor record his last few months in office. We agreed to mediation, ate crow, and agreed to hire a labor counselor at NASA Headquarters to keep us from future labor errors. However, government workers did complete the construction of the Saturn I launch complex, the one that President Kennedy would later visit during his last week in office.

Eisenhower and Lunar Exploration

After Kennedy's election, President Eisenhower held a cabinet meeting on 20 December, and space exploration was on the docket. Keith went first and discussed the NASA fiscal year (FY) 1962 budget submission to Congress. Little discussion followed.

---

2. Jane Van Nimmen and Leonard C. Bruno, with Robert L. Rosholt, Table 4.11, "Funding NASA's Program FY 1962," in *NASA Historical Data Book, Volume I: NASA Resources 1958–1968* (Washington, DC: NASA SP-4012, 1988), p. 138.

Then Dr. Kistiakowski, the President's science advisor, followed with a presentation of his committee's study on making a manned lunar landing. All were attentive. When he said it was difficult to determine costs, heads nodded. But he went on to say that estimates ranged from $26 to $38 billion. The room was filled with sighs, and someone volunteered, "If we let scientists explore the Moon, then before you know it they'll want funds to explore the planets." Everybody laughed. Eisenhower ended this part of his meeting with a rhetorical question: "Can anybody tell me what is the best space program for $1 billion?" Walking from the cabinet room, I realized why Maury Stans was adamant that there would be no additions to NASA's budget in FY 1962.

Space Exploration Council

On 5 January 1961, the Space Exploration Council held a full-day session to discuss a program for manned lunar landing. George Low, Program Chief for Manned Space Flight, introduced the subject by outlining the guidelines of the program. His stated objective was a lunar landing and safe return at the earliest practical date, regardless of cost. The establishment of a lunar base was the secondary goal. In his view, consideration should be given to using a number of Saturn launch vehicles with rendezvous in Earth orbit, as well as to a direct approach with a single Nova-type vehicle (a vehicle capable of both a manned lunar landing and a safe return). He recommended holding the schedule for the Saturn I unchanged but changing the Saturn II's first flight from July 1965 to April 1964. In his study he assumed a spacecraft weight of 8,000 pounds.[3]

Following Low's presentation, Wernher von Braun outlined Marshall Space Flight Center's plans, which were based on more modest funding. He stated that the lunar program should do the following things:

- Use building blocks from the present spaceflight program

- Possess flexibility in case of technical mishaps or breakthroughs

- Be adaptable for rapid expansion if the need arises

- Fit into the time and economy framework of the nation

- Be attractive to the general public and the military[4]

Then Max Faget, representing the Space Task Group (which became the Manned Spacecraft Center in Houston), stated Apollo's objectives:

- On-board capability to maneuver in deep space

- Ability to perform rendezvous missions

- Capability for outer space (hyperbolic) reentry with landing at a predetermined location

- Ability to terminate at any time with safe crew return.[5]

The presentations were not coordinated prior to the meeting. There were a wide variety of schedules presented, and the conference room was awash with billion-dollar estimates. There was certainty on one issue: NASA's leadership had taken a giant intellectual step since the industry conference of July 1961. Then, NASA's planning goal for the decade, based on the earlier Goett Study (chaired by Harry Goett, Director of Goddard Space Flight Center), was circumlunar flight. The Goett Committee felt that there would be too many imponderables in a manned lunar landing to warrant further investigation in the near term. However, now there was clear consensus that NASA should proceed with the lunar landing planning and that George Low should be its chief honcho. Before the meeting ended, Keith Glennan warned that Eisenhower hadn't approved the mission. His admonishment was certainly an understatement. But for Keith,

3. George M. Low, presentation to Space Exploration Council, 15 January 1961, in *A Program for Manned Lunar Landing*, folder 7020, NASA Space Exploration Program Council (SEPC), NASA Headquarters Historical Reference Collection, Washington, DC.

4. George C. Marshall Space Flight Center, presentation to the NASA Space Exploration Council, 5 January 1961, in *Lunar Transportation Systems*, folder 7020, NASA SEPC, NASA Headquarters Historical Reference Collection, Washington, DC.

5. Max Faget, George C. Marshall Space Flight Center, presentation to the NASA Space Exploration Council, undated, folder 7020, NASA SEPC, NASA Headquarters Historical Reference Collection, Washington, DC.

President Eisenhower would have recommended to Congress that no further manned space mission should be in the works until Mercury was completed and evaluated.

The instructions for the manned lunar landing task group under George Low's direction were dated 6 January 1961. The principal items requested of the group follow:

> It is the task of this group to prepare a position paper for use in presenting the NASA FY62 budget to Congress. The paper should answer the question "What is NASA's Manned Lunar Landing Program?"

> The Program for FY62 is defined in the budget for FY62 and in our plans for the conduct of the program utilizing these funds. The task group must put these individual pieces together into a complete but tersely worded statement of the NASA Lunar Program for FY62.

> Since a single year's program cannot stand alone it is obvious that the Congress will be interested in what we plan to accomplish in the following years. This information is summarized in the Ten Year Plan. We do not have enough data to decide at this time whether we will attempt manned landing by direct flight or by rendezvous techniques.

> Finally, the paper must answer the question, "How much is it going to cost to land a man on the moon and how long is it going to take?" We must answer this question for both the rendezvous and the direct approach.[6]

Abe Silverstein, Director of the Office of Space Flight Programs, and I attended the first meeting of the Lunar Landing group on 9 January. Questions arose and were clarified. A summary of those is listed below:

- We must not assume that a decision has been made to land a man on the moon.

- However, development of the scientific and technical capability for manned lunar landing is a prime NASA goal but it is not *the* only goal.

- In paragraph 5 of the January 6 instructions it is not intended that we develop specific dates and costs. This is not possible at this time. The position paper must spell out what our answer should be to the question.

- We must present a positive rendezvous program. This program will be pursued in order to develop a manned spacecraft capability in near space, regardless of whether it is needed for manned lunar landing.

- Our approach should be positive. We should state that we are doing the things that must be done to determine whether manned lunar landing is possible.[7]

Keith Glennan's Last Day

Jack Kennedy's inauguration was on 20 January; since Keith Glennan would be leaving NASA as Eisenhower left office, he had to wrap up his affairs at NASA on the 19th. He had a busy day and put the capstone on much unfinished business. One such item was Ranger, along with Surveyor; both were handled by the Jet Propulsion Laboratory, NASA's Center for unmanned lunar and planetary missions. Ranger, a lunar photographic probe, was already under development. Photographs were to be transmitted from Ranger as it approached and crashed on the lunar surface. Surveyor's role was quite the opposite; it was to land softly on the Moon and analyze surface conditions after impact. By 19 January, the source evaluations were ready for the Administrator's presentation. Keith gamely held off his return to Cleveland for the source selection. Hughes Aircraft won the Surveyor contract. The data from Surveyor would be crucial to the design of the manned Lunar Lander. During the day, Keith also documented

---

6. George Low, "Instructions to Manned Lunar Landing Task Group," 6 January 1961, folder 7020, NASA SEPC, NASA Headquarters Historical Reference Collection, Washington, DC.

7. George Low, "Further Instructions to the Manned Lunar Landing Task Group," 9 January 1961, folder 7020, NASA SEPC, NASA Headquarters Historical Reference Collection, Washington, DC.

those projects he had authorized.[8] In each case, he listed limitations, requirements, and understandings relating to technical parameters experiments and management, as well as magnitude and type of resource allocation. Among the projects were 16 scientific satellites and probes, 2 meteorological satellites, 3 nonactive communication satellites, 7 lunar and planetary missions, 2 manned spacecraft, 4 launch vehicle developments, 2 rocket engine developments, and 5 nuclear projects for power or propulsion. The two rocket engines were the F-1, which was kerosene-fueled with a thrust of 1.5 million pounds, and the J-2, hydrogen-fueled with a thrust of 200,000 pounds. These engines were central to the success of the Saturn vehicles. Of course, the great success of Mercury in the Kennedy years was due to the planning and product development in Eisenhower's administration. In two and a half years, NASA was up and away with a space program that provided a solid foundation for the years to come.

Keith was due for a good change of pace, but it wouldn't start for at least 24 hours. After a glass of sherry to toast his performance, Keith left for his apartment and then the drive home to Ohio. Unfortunately, there was a blizzard of major proportions. Keith reached his apartment, gathered up his remaining luggage, and started driving. After struggling for a few hours and gaining only a few miles, he headed to a friend's house for emergency lodging. He then returned home to family, friends, and his beloved Case Institute the following day.

The Wiesner Ad Hoc Committee on Missiles and Space

During the interval between Kennedy's election and his inauguration, a sword of Damocles hung over NASA. Jerry Wiesner chaired the incoming administration's committee on missiles and space. Alarming rumors, which we thought were probably inaccurate, kept appearing in journals and newspapers. Such ideas as a merger of NASA and the military or a transfer of manned spaceflight to the military, along with hints about the incompetence of NASA leadership, were quite unnerving. The actual report by the ad hoc Committee on Space, dated 10 January 1961 (appearing 10 days before the inauguration) was fairly reasonable, although I bristled a bit at the time.

The report noted, quite rightly, that space exploration had captured the imagination of the peoples of the world. It was important to maintain American preeminence in space—the prestige of the United States was on the line. The report again correctly pointed out that the inability of U.S. rockets to lift large payloads into space seriously limited our program. But then, in the section on Man-in-Space, the report stated that by placing a high priority on the Mercury Project, we had strengthened the popular view of its importance as compared with the "acquisition of knowledge and the enrichment of human life."[9] It's true that the public became more excited by the selection of our astronauts than by Dr. Van Allen's discovery of the radiation belts around Earth, but that was caused more by the human interest than by the contents of NASA's public releases.

The report then expressed great concern about the possible failure of Mercury and the resulting possible loss of life. The new administration would have to take the blame for the death of an astronaut. The report went on to say that the Man-in-Space program appeared unsound and that the new administration should be prepared to modify it drastically or cancel it. Above all, it recommended that Mercury be downgraded and project advertising stopped.

The report went on to say that the difficulties and delays endured by the program had resulted from insufficient planning and direction caused by a lack of "a strong scientific personality in the top echelons."[10] Not only had this lack affected NASA's operations, but there were also far too few outstanding scientists and engineers deeply committed to the space field in general. Strengthening NASA's

---

8. T. Keith Glennan, "Authorized Development Projects," 19 January 1961 memorandum, Robert Channing Seamans, Jr., papers, MC 247, Institute Archives and Special Collections, MIT Libraries, Cambridge, MA.

9. Wiesner Committee, "Report to the President-Elect of the Ad Hoc Committee on Space," 10 January 1961, reprinted in *Exploring the Unknown: Selected Documents in the History of the U.S. Civil Space Program Volume I: Organizing for Exploration*, ed. John M. Logsdon, Linda J. Lear, Jannelle Warren-Findley, Ray A. Williamson, and Dwayne A. Day (Washington, DC: NASA SP-4407, 1995), p. 422.

10. Ibid., p. 421.

top management would encourage more talented personnel to participate.

However, in the same report, there was the already-mentioned acknowledgment that the United States was operating at a disadvantage because our boosters had limited capability compared to those of the Soviets. The Saturn booster was endorsed, along with the Centaur rocket and the F-1 engine—all part of Glennan's legacy. The report had another strong plug for the past scientific: "In the three years since space exploration began, experiments with satellites and deep space probes have provided a wealth of new scientific results of great significance. In spite of the limitations in our capability of lifting heavy payloads, we now hold a position of leadership in space science."[11] Not too bad for a bunch of dimwits!

Finally, the report laid out application possibilities for communication, meteorology, and further scientific investigation in keeping with NASA's existing plans. It stressed the need for wider participation by university and industrial scientists. So NASA's number-one issue in the Kennedy administration was going to be "where goeth man in space?"[12]

During this period of anxiety, there was much excitement as the inaugural activities went into high gear. A blizzard made it difficult to get to evening events the night before. Our daughter was undaunted, walking out the front door of our house in an evening gown with appropriate slippers and no overshoes. We arrived late at Constitution Hall for the concert, minutes after the President-elect's departure. My parents arrived at 4:00 a.m. By chance, they were on a plane from Boston with Cardinal Cushing, who was officiating at the swearing in and whose entourage included 45 nuns. When landing in Washington became impossible, they were diverted to New York and took a train to Washington. The day itself was sunny and cold, and an exuberant crowd was full of confidence in the new leadership.

11. Ibid., p. 420.
12. Ibid., p. 420.

Chapter 3:
THE KENNEDY CHALLENGE

Ham Gets a Sporty Ride

Eleven days after the inauguration, Ham, a chimpanzee, was strapped down in Mercury Redstone (MR-2), ready for liftoff. The first launch of Mercury had occurred on the 19th of the previous December. The mission was unmanned and used a Redstone launch vehicle and a boiler-plate capsule. The results were sufficiently successful for a chimpanzee but not a human to board MR-2. Six chimpanzees were at the Cape, accompanied by 20 medical specialists and animal handlers from Holloman Air Force Base. At liftoff, Ham was pronounced stable, working his levers perfectly to avoid the punishment that came from inattention. At waist level, there was a dashboard with two lights and two levers. Ham knew well how to stay comfortable by avoiding the electrical shocks that followed errors. Each operation of his right-hand lever, cued by a white light, postponed a shock for 15 seconds. At the same time, Ham had to press a left-hand lever within 5 seconds of the flashing of a blue light every 2 minutes. During the flight, Ham achieved a perfect score with his left hand and made only two mistakes out of 50 prompts with his right. He did receive two mild shocks for his mistakes, but he also received banana pellets for his left-handed performances. The cockpit photos showed a surprising amount of dust and debris during weightlessness.

The Redstone Launch vehicle accelerated the capsule to too high a velocity at cutoff (5,857 miles per hour instead of 4,400 mph), so Ham experienced 14.7 g's rather than 12 g's on reentry, and he landed in the Atlantic 132 miles beyond the planned impact point. Because of leaks in the capsule, the capsule had 800 pounds of water at pickup. However, when deposited on the USS *Donner*, Ham appeared in good condition and readily ate an apple and half an orange. Could human beings have done as well?[1]

A manned lunar landing task group was established on 6 January 1961 as a result of the many questions that arose at the Exploration Council as noted in chapter 2. The report by the Manned Lunar Landing group was submitted to the Associate Administrator on 7 February 1961. The findings of this group were remarkably prescient and most important to NASA in the months that followed. The group found that no inventions or breakthroughs were believed to be required to ensure safe manned lunar flight. It went on to say that booster capability could be acquired either by a number of Saturn C-2 launches followed by rendezvous and docking or by Nova, a launch vehicle larger than the Saturn. The group found that rendezvous techniques could allow a lunar landing in significantly less time than the other two options.

The group's report stated that Mercury would have most of the on-board systems required in the future. They expected that many of the systems for lunar landing would be outgrowths of this effort. The need for special guidance and navigation in lunar approach, orbit, and landing was omitted by the group members, but they did stress the importance of the F-1, J-2, and RL10 rocket engines for the development of the Saturn and Nova launch vehicles. From a biological standpoint, the group recommended that studies be accelerated on the effects of weightlessness and radiation. It noted that these environmental conditions would become increasingly important as astronauts extended their time in orbit and as missions moved farther from Earth and the protective shielding of Earth's atmosphere and the magnetically induced Van Allen radiation belts.

The Apollo A using the Saturn C-1 would allow multimanned orbital flights in 1965. The advanced, long-duration Apollo B launched by the Saturn C-2 would provide the capability for circumlunar and lunar orbital missions in 1967.

The group felt that the manned lunar landing could occur as early as 1968 and as late as 1971. Whether it would be early or late hinged on the via-

bility of rendezvous operations. Rendezvous operations obviated the need for the super booster called Nova, which the group estimated would require an extra one to two years. Hence, the manned lunar landing was bracketed between 1968 and 1969 when using rendezvous maneuvers, or between 1970 and 1971 if direct ascent with a single launch vehicle was the chosen mode. The mission, spacecraft, launch vehicle, and dates are shown in figure 1. Fortunately, Nova was not required.

The cost estimates were low, with $3 billion for the spacecraft and $4 billion for the launch vehicle—a total of $7 billion. However, much was omitted, including the Gemini missions, and the estimated cost of facilities and operations was considerably less than what was actually required. Notwithstanding, the report by George Low and his group was most valuable in the meetings with the President and Congress that were soon to follow.

James E. Webb Takes Charge of NASA

James E. Webb was nominated as the Administrator of NASA in early February 1961, and needless to say, I was most anxious for a meeting in order to find out whether I would soon be departing. At our first discussion, he emphasized leadership and asked my views on the effectiveness of Sears Roebuck's dispersed management versus Montgomery Ward's hierarchical organization. Fortunately, it was a subject I'd studied at Columbia's advanced management program the previous summer, so I felt pretty comfortable in my exchange of ideas. Jim asked both Hugh Dryden and me to remain at NASA, and over time, we became known as the Triad—each of us had different skills and responsibilities, but we convened (figure 2) to make key decisions that were usually unanimous.

Jim was sworn in on 12 February 1961, and, soon thereafter, a meeting was arranged with the new Director for the Bureau of the Budget, Dave Bell. The previous administration had reduced our budget by $300 million, so we decided to request an additional $190 million for manned-flight-related projects and $10 million for communication satel-

---

1. Loyd S. Swenson, Jr., James M. Grimwood, and Charles C. Alexander, *This New Ocean: A History of Project Mercury* (Washington, DC: NASA History Series, 1989), p. 310.

Manned Missions and Launch Dates

MISSION	SPACECRAFT	LAUNCH VEHICLE	DATE
EARTH ORBITING 1 Man, Short Duration	*MERCURY*	*ATLAS*	1961
EARTH ORBITING 3 Men, Long Duration	APOLLO "A"	SATURN C-1	1965
CIRCUMLUNAR, LUNAR ORBIT 3 Men	APOLLO "B"	SATURN C-2	1967
MANNED LUNAR LANDING Orbital Operations Direct Approach	APOLLO "B" APOLLO "B"	SATURN C-2 NOVA	1968-69 1970-71

Figure 1. Results of a study commissioned on 6 January 1961 and chaired by George Low. These findings were available on 7 February 1961.

lites. Dave Bell told us that the President was most interested in space exploration and planned to get his mind around the issues in connection with the next fiscal year's budget, that of FY 1963. Mr. Webb demurred, saying that the issues couldn't wait, and so a session took place with the President, the Vice President, their staff, and the Director of BoB on 22 March.

First Meeting with President Kennedy

As was the custom, the Director of BoB started the meeting by advising the President that additional funding should await the review of the following year's requirements. Mr. Webb then said that I would present NASA's request. The President asked how long it would take; when Jim responded that it would be 30 minutes, the President said that he had only 15. The phone then rang, and the President had an extended conversation with the Speaker of the House. Ultimately, I had an opportunity to summarize our recommendations. The President looked at me and said, "That was very

good; I would like your views in writing tomorrow." I wrote the memo that evening, hand-delivered it to Jim Webb the following morning, and then joined my family in Mt. Tremblant, Canada, for a weekend of skiing. The memo was forwarded by Jim Webb and contained these requests:

The funding rates of five projects were discussed at the NASA-BoB conference with the Vice President and the President on March 22, 1961. An agenda prepared prior to the meeting summarized the objectives of these projects and indicated in each case the effect of the funding rate on the schedule. The multi-manned orbital laboratory is contingent upon the Saturn C-1 which is adequately funded, and a new spacecraft for which NASA recommends an increase from $29.5 to $77.2 million. This increase starts an accelerated program leading to multi-manned orbital flights in 1965 rather than 1967.

The multi-manned circumlunar flight requires the Saturn C-2 and a spacecraft which will evolve from the design of the

Figure 2. NASA Management Triad in the office of James E. Webb (center). He and Dr. Robert C. Seamans, Jr. (right), listen as Dr. Hugh Dryden (left) has the floor. (NASA Image Number 66-H-93)

orbital spacecraft. The recommended $73 million increase in FY 1962 funding for the Saturn C-2 leads to the completion of the Saturn development in 1966, and manned circumlunar flight in 1967 rather than 1969.

A manned lunar landing requires a new launch vehicle with capabilities beyond Saturn. This vehicle, called Nova, is still under study. It would use a first-stage cluster of the 1.5 million pound thrust, chemically fueled engines, which we have under development. We are requesting $10.3 million additional over the present FY1962 budget to accelerate the engine development. The first manned lunar landing depends upon this chemical engine as well as on the orbital and circumlunar programs and can be achieved in 1970 rather than 1973.[2]

Notice that the dates in this memo were consistent with those in George Low's working group on lunar landing. Also included in the request was Centaur, which, with Atlas as the first stage, would send unmanned probes to soft-land on the Moon. The Centaur RL10 liquid-hydrogen engines were also to be used in the Saturn I upper stage. Of the total $200 million requested, the President decided to support communication satellites with $10 million and propulsion projects with $115.7 million, but the money would not support the multimanned orbiting laboratory.[3]

A New Ball Game

Sergey Korolev was the prime mover of the Soviet space program from its inception until his death in 1966. Originally an aeronautical engineer, he was imprisoned in the late 1930s after being accused of sabotage. Stalin, not noted for his receptivity to challenging ideas, banished Korolev to a forced labor camp in Siberia, where he languished until the Soviets were desperate for engineers in World War II. A special camp was established just outside Moscow, and Korolev was moved there. He

performed so well that he was eventually released. At the war's end, he was sent to Peenemünde to obtain engineers, technical information, and equipment related to the German V-2 development. Later, he convinced Chairman Khrushchev to support a few satellite launches using the Soviet ballistic missile program. Sputnik was an instant success that opened the way for Korolev and his team to embark on a broad-scale space endeavor. Korolev struck again on 12 April 1961 (see figure 3), when Yuri Gagarin orbited Earth and landed safely to tremendous acclaim in the Soviet Union and around the world. Our Congress went berserk, and President

Figure 3. Sergey P. Korolev, founder of the Soviet space program, shown here in July 1954 with a dog that had just returned to Earth after a lob to an altitude of 100 kilometers on an R-1d rocket. (Source: http://grin.hq.nasa.gov/ABSTRACTS/GPN-2002-000163.html)

---

2. Robert C. Seamans to James E. Webb, 23 March 1961, Robert Channing Seamans, Jr., papers, MC 247, Institute Archives and Special Collections, MIT Libraries, Cambridge, MA.

3. Table 4.13, "Funding NASA Program in FY1962," in *NASA Historical Data Book*, p. 138.

Kennedy was distressed. The following day, Mr. Webb and Dr. Dryden were called before the House Authorization Committee on Space and Aeronautics in the Caucus room. Jim and Hugh were pressed for bolder action and parried the thrust of the committee members in admirable fashion. The day after, it was back to the Manned Space Subcommittee for George Low and myself. The hearing was held in the old committee chambers. George began his testimony but was interrupted by Congressman David King of Utah:

MR. KING: May I make a comment there and then, and then, if you will, carry on. I understand the Russians have indicated at various times that their goal is to get a man on the Moon and return safely by 1967, the 50th anniversary of the Bolshevik Revolution. Now specifically I would like to know, yes or no, are we making that a specific target date to try to equal or surpass their achievement?

DR. SEAMANS: As I indicated in earlier testimony this morning, our dates are for a circumlunar flight in 1967 and a target date for the manned lunar landing in 1969 or 1970.

MR. KING: That of course—then that outlines the issue very squarely. As things are now programmed we have lost. The score will be three to nothing for the Russians. I would like to make it clear for the record that I personally—and I am not a technical man, I am speaking just as a Congressman, trying to do what I can for the country—that I would favor any such program, regardless of the cost, that would put us definitely in the race to reach the Moon first. I think anything short of that will be doing an injustice to our country. Let me just ask this final question. Do you think it would be conceivably possible, by increasing appropriations, by marshaling our manpower and resources and everything else we have available, to meet this target date of, let us say, 1967?

DR. SEAMANS: This is really a very major undertaking. To compress the program by 3 years means that greatly increased funding would be required for the interval of time between now and 1967. I certainly cannot state that this is an impossible objective. If it comes down to a matter of national policy, I would be the first to review it wholeheartedly and see what it would take to do the job. My estimate at this moment is that the goal may very well be achievable. That is the best answer I can give you at present.

MR. KING: I think that is a very significant statement and I am very grateful to get it . . . .[4]

There followed an exchange with a Republican member of the committee (J. Edgar Chenoweth of Colorado) and a final question by the committee's chairman, Congressman George Miller (Democrat):

MR. CHENOWETH: I understood from your last answer to Mr. King that you thought it could be done. That impression will go out. I think you have to be very careful what you tell this committee because there will be those who will say, "All right, lets boost up our appropriation, double it, treble it. The most important thing is to put a man on the Moon." I don't know that it is. I doubt it. But some feel that way. I think it is a high policy decision to be made and to be made shortly. I think it is important you word your answers carefully here, because the wrong interpretations may be placed upon them not only by this committee but by those who will read stories that will go out.

DR. SEAMANS: I disagree on one point you touched upon earlier. I feel this committee is a most important forum for discussion of this issue. I believe there are other important forums. I agree this is a most important national issue.

MR. CHENOWETH: The question is whether it is of such great importance that we can afford to neglect other programs that perhaps may involve a change of our whole fiscal program in order to accomplish this one objective. Is it that important, in your opinion?

DR. SEAMANS: Obviously I cannot answer that question.

4. House Committee on Science and Astronautics and Subcommittees Nos. 1, 3, and 4, Hearings, 87th Cong., 14 April 1961.

MR. CHENOWETH: It is a decision to be made at a higher level.

DR. SEAMANS: I think it is a decision to be made by the people of the United States.

MR. CHENOWETH: How will they make it?

DR. SEAMANS: Through the Congress and through the President. It is a matter of national importance to have specific objectives for our space effort.

MR. CHENOWETH: I disagree. The people of this country do not have the technical knowledge on this subject that you have. When you talk about placing a man on the Moon, they don't know what you are talking about. They don't know what expenditure is involved, nor the scientific and research work that has to be done. We can't expect them to make that decision.

MR. MILLER: Is this not our responsibility as the representatives of the people . . . .[5]

When the hearing was over, George Low and I faced a barrage of reporters and a battery of TV cameras as we left the building. I felt there might be a concern about my performance and headed directly to Mr. Webb's office, where Nina Scrivener, his secretary, listened thoughtfully to my message: "Tell Mr. Webb I did the best I could, but the White House may be quite unhappy." I knew it was unwise for an underling to get out ahead of the President. I found out later that Ken O'Donnell, the President's political advisor, wrote a strongly worded letter to Mr. Webb about my performance, but in his return letter dated 21 April, Jim supported me. He noted, "My judgment from the record and my personal experience with the committee is that our group, particularly Dr. Seamans has done a splendid job for this administration. Dr. Seamans bore the brunt of discussions as to our relations with the Bureau of the Budget and the President. From a reading of the testimony I believe Seamans has done an exceptionally fine job."[6] Keith Glennan wasn't so kind. He wrote, "I think an unfortunate state-ment by Bob Seamans before a congressional committee gave the newspapers and through them the public, the idea that this flight [lunar landing] was to be accomplished by late 1967."

A Call to the Vice President

On 20 April, President Kennedy wrote Vice President Johnson a memorandum in which he asked:

1. Do we have a chance of beating the Soviets by putting a laboratory in space, or by a trip around the moon, or by a rocket to land on the moon, or by a rocket to go to the moon and back with a man? Is there any other space program which promises dramatic results in which we could win?

2. How much additional would it cost?

3. Are we working 24 hours a day on existing programs? If no, why not? If not, will you make recommendations to me as to how work can be speeded up.

4. In building large boosters should we put our emphasis on nuclear, chemical, or liquid fuel, or a combination of these three?

5. Are we making maximum effort? Are we achieving necessary results?

I have asked Jim Webb, Dr. Wiesner, Secretary McNamara and other responsible officials to cooperate with you fully. I would appreciate a report on this at the earliest possible moment.[7]

The Whirlwind Week of 2 May 1961

The week started with reasonable assurance that in a few days, NASA was going to be tested in the eyes of the world by Alan Shepard's Mercury flight. And then, if that was successful, NASA was going to embark on a lunar program even before

---

5. House Committee on Science and Astronautics and Subcommittees Nos. 1, 3, and 4, Hearings, 87th Cong., 14 April 1961.

6. James Webb to Ken O'Donnell, 21 April 1961, Robert Channing Seamans, Jr., papers, MC 247, Institute Archives and Special Collections, MIT Libraries, Cambridge, MA.

7. President John F. Kennedy to Vice President Lyndon Johnson, 20 April 1961, Robert Channing Seamans, Jr., papers, MC 247, Institute Archives and Special Collections, MIT Libraries, Cambridge, MA.

the United States had sent an astronaut to orbit Earth. If that happened, there would be a clear need for an in-depth investigation of all the steps that would have to be taken and of the costs and time that would be involved. On 2 May, I sent a memorandum to the directors of the four program offices establishing an ad hoc task group for this study. Bill Fleming, my program assistant, was to head the study, and the individuals assigned to the study were to be on a full-time basis for the duration of the effort.

Friday, 5 May 1961, Mercury Redstone (MR-3), Alan Shepard

Later in the week, all eyes were on Alan Shepard at Cape Canaveral. Jerry Wiesner, in his interregnum report, had alerted the Kennedy administration that if they weren't careful, they'd own the Mercury project. The hour of truth had arrived. Should NASA be allowed to launch the MR3 with Alan Shepard aboard? The mission had been carefully and responsibly reviewed by a White House committee chaired by Donald Hornig. His committee was favorably impressed with NASA's planning and testing. But supposing the launch was a disaster, especially following Gagarin's achievement? Ed Welsh, secretary of the Space Council, joined me on Friday, 5 May, to follow the mission on an in-house circuit. At that time, there was small, obscure room in NASA Headquarters, across from the White House, where the voice of the Mission Director was piped in. Ed confirmed that there was much concern about possible failure, but he had raised the question, what if we succeed? Anyway, it was now a "go." Hugh Dryden was at the Cape as NASA's senior observer. He had been close to the Mercury program since inception and was clearly the person to have on hand in the event of unexpected contingencies.

Freedom 7 roared off at 10:34 and started its climb. The ride was smooth and the voice communication clear for the first 45 seconds. Buffeting started in the transonic zone and became severe about 90 seconds into the flight at maximum dynamic pressure. Alan's head was bouncing so hard that he couldn't read the flight instruments.

The maximum g forces occurred after 2 minutes, and the engines cut off 22 seconds later. Alan was traveling 5,134 mph, the desired speed. He had been traveling face-forward when, at 3 minutes into the flight, the capsule automatically turned completely around in preparation for reentry. Now it was time for the most important task, determining whether a human could control the capsule. He switched onto manual control one axis at a time. He first used his right grip backward to tilt his heatshield downward 34 degrees for reentry. Later, he was pleased to find that he could control the spacecraft's movement about all three axes—roll, pitch, and yaw—and the fuel use was similar to what he had experienced with the Mercury trainer. When the retrorockets fired at the appropriate time, it provided what astronauts later described as a "comforting kick in the ass."

As *Freedom 7* approached the atmosphere, the 0.05-g light came on, and the acceleration rapidly built up to a peak of 11.6 g's. As the spacecraft entered the atmosphere, the drogue chute first opened at 21,000 feet; the main chute followed at 10,000. The recovery forces were standing by for pickup. Alan felt that the thud at impact was comparable to that of a carrier landing. After landing, the chutes were released, with the capsule listing 60 degrees to starboard. The rescue helicopter was soon overhead, and Alan was taken aboard the carrier *Lake Champlain* 11 minutes after landing. Ed Welsh and I did a few war whoops in our cubicle, shook hands, and gave thanks for all those involved in the flight's success.

Upon examination, doctors found that Alan had suffered no ill effects, and, as he reported himself, weightlessness was "quite pleasant." A half hour into his free-dictation report, Alan was summoned to the bridge deck for a call from President Kennedy. Kennedy had followed the flight closely via television and was now offering his congratulations.

There was mostly worldwide acclaim, but chagrin in Moscow, where Premier Nikita Khrushchev asked why the "up and down" flight of Shepard gained such extensive media publicity even though Gagarin had long since orbited the world.[8]

---

8. Swenson, Grimwood, and Alexander, *This New Ocean*, pp. 352–357.

Big Doings at the Pentagon

On Saturday, Hugh Dryden was still at Cape Canaveral awaiting Alan's arrival and debriefing. Jim Webb, Abraham Hyatt (Director of Plans and Program Evaluation), and I arrived in Bob McNamara's office at the Pentagon. Bob had Roswell Gilpatric, his deputy, and John Rubel, head of space research and development in the Department of Defense (DOD), with him. The Vice President had turned to NASA and DOD to help answer the President's request for recommendations on U.S. space policy and direction. The Vice President said that NASA and DOD would have most of the action, so the administration needed our views on whether there was any space program that promised dramatic results that we could achieve before the Soviets.

McNamara greeted us crisply. Once seated, he suggested that we lay our cards on the table, and he asked Jim to go first. As per our plan, Jim first recommended that NASA proceed with a manned lunar landing mission. It was our view that the Soviets could conduct a manned orbital laboratory or a circumlunar mission with means already available. However, McNamara questioned our views and suggested a planetary trip to Mars. I found his suggestion horrifying and pointed out that we had neither the technology nor the physiological understanding to proceed with such a mission. The discussion recognized the previous day's achievement by Alan Shepard and noted that the highly favorable media response resulted from the mission's being carried out completely in the open. It had become obvious that national prestige should be recognized as one of four valid reasons for space undertakings. The other three reasons were scientific investigation, commercial value, and national security. From this meeting resulted a report to the Vice President that recommended a $626-million add-on for FY 1962, of which $549 million was for NASA.[9] The line items for NASA funding included the following:

- Apollo for multimanned orbital laboratory

- Nova, a large launch vehicle, for manned lunar landing

- Scientific experiments in space

- Satellite communications

- Meteorological satellites

- Nuclear rocket developments

The major share of the funding recommendation was earmarked for Apollo and Nova. To quote from the report:

To achieve the goal of landing [a person] on the moon and returning him to earth in the latter part of the current decade requires immediate initiation of an accelerated program of spacecraft development. The program designated Project Apollo includes initial flights of a multi-manned orbiting laboratory to qualify the spacecraft and manned flights around the moon before attempting the difficult lunar landing.

The advanced goal of manned landing on the moon also requires the development of a launch vehicle (Nova) with a first stage thrust of about six times that of the largest vehicle now under development (Saturn I) [Nova was never started; however the Saturn V had nearly five times the thrust of the Saturn I under development].[10]

In addition to the specifics in the report, there was a general section on the need for close cooperation and coordination between NASA and DOD. In particular, the report noted the importance of the manned lunar landing in the context of a total national effort.

The future of our efforts in space is going to depend on much more than this year's appropriations or tomorrow's new idea. It is going to depend in large measure upon the extent to which this country is able to establish and to direct an integrated national space program. To quote further from the report:

We recommend that our National Space Plan include the objective of manned lunar explo-

---

9. Table 4.13, *NASA Historical Data Book*, p. 138.

10. James E. Webb and Robert McNamara, "Recommendations for Our National Space Program: Changes, Policies, Goals," report to Vice President Lyndon B. Johnson, 8 May 1961.

ration before the end of this decade. It is our belief that manned exploration to the vicinity of and on the surface of the moon represents a major area in which international competition for achievement in space will be conducted. The orbiting of machines is not the same as the orbiting or landing of man. It is man, not merely machines, in space that capture the imagination of the world . . . .

The establishment of this major objective has many implications. It will cost a great deal of money. It will require large efforts for a long time. It requires parallel and supporting undertakings which are also costly and complex. Thus for example, the RANGER and SURVEYOR unmanned probes and the technology associated with them must be undertaken and must succeed to provide the data, the techniques and the experience without which manned lunar exploration cannot be undertaken.

The Soviets have announced lunar landing as a major objective of their program. They may have begun to plan for such an effort years ago. They may have undertaken important first steps which we have not begun.

It may be argued, therefore, that we undertake such an objective with several strikes against us. We cannot avoid announcing not only our general goals but many of our specific plans, and our successes and failures along the way. Our cards are and will be face up—theirs are face down.

Despite these considerations we recommend proceeding toward this objective. We are uncertain of Soviet intentions, plans or status. Their plans, whatever they may be, are not more certain of success than ours. Just as we accelerated our ICBM[11] program we have accelerated and are passing the Soviets in important areas in space technology. If we set our sights on this difficult objective we may surpass them here as well. Accepting the goal gives us a chance. Finally, even if the Soviets get there first, as they may, and as

some think they will, it is better for us to get there second than not at all. In any event we will have mastered the technology. If we fail to accept this challenge it may be interpreted as a lack of national vigor and capacity to respond.[12]

The DOD had already prepared a draft report for submission to the Vice President. John Rubel and I were given the job of editing the report and bringing it into concert with the Saturday meeting. We worked together well into the evening, when Jim Webb arrived after escorting Alan Shepard's parents to their hotel. Jim, John, and I completed the final editing at 2:00 Monday morning. John and I reviewed the retyped copy and brought it to McNamara and Webb for signature on Monday morning, prior to the 9:00 a.m. ceremony at the White House honoring Alan Shepard.

A Hero's Welcome

Following receipt of his honors at the White House (see figure 4), Alan Shepard was sped in a motorcade to the Capitol, where he addressed a joint session of Congress. There followed a special reception and luncheon, hosted by Vice President Johnson, at the State Department. Near the end of lunch, the Vice President stood to toast Alan and his family and then left to meet with the President before leaving for Vietnam. In his hand was the envelope containing the McNamara-Webb report completed earlier that morning.

A Special Message to Congress

At this juncture in the space program, it is interesting to compare the derivation of the USSR and U.S. programs. Both derived considerable strength from the German effort at Peenemünde, the USSR from Korolev's hiring of technical personnel and collectors of data and hardware and the United States from the capture of Dr. von Braun and his management team. The von Braun team became the Army's Ballistic Missile Agency of the Redstone Arsenal prior to its transfer to NASA. Other ingredients transferred to NASA were the laboratories of

---

11. Intercontinental ballistic missile.

12. Webb and McNamara.

Figure 4. President John F. Kennedy congratulates astronaut Alan B. Shepard, Jr., the first American in space, on his historic 5 May 1961 ride in the Freedom 7 *spacecraft and presents him with the NASA Distinguished Service Award. (NASA Image Number 1961ADM-13, also available at* http://grin.hq.nasa.gov/ABSTRACTS/GPN-2000-001659.html*)*

the NACA, the Jet Propulsion Laboratory of the California Institute of Technology, and the Navy's Vanguard team. Figure 5 shows how these diverse groups had coalesced by 1961.

Under Korolev, the Soviets had orbited the satellite Sputnik, a dog, and the cosmonaut Yuri Gagarin. They had also photographed the far side of the Moon. The United States had launched Explorer, a weather satellite, the Echo balloon, and Alan Shepard into suborbital flight.

Several days prior to 24 May, when President Kennedy was to address a joint session of Congress, Jim Webb received a copy of that part of the speech related to space. Sure enough, the President was recommending a manned lunar landing and safe return, but in 1967. Jim called Ted Sorensen, the President's speechwriter, to request a change of date. The country should operate in the open, he said, but shouldn't make such a bold commitment in terms of time. The compromise with the White

House was "within the decade." Excerpts from the President's speech follow:

> Since early in my term, our efforts in space have been under review. With the advice of the Vice President, who is Chairman of the National Space Council, we have examined where we are strong and where we are not. Now it is the time to take longer strides—time for a great new American enterprise—time for this nation to take a clearly leading role in space achievement, which in many ways may hold the key to our future on earth.
>
> Recognizing the head start obtained by the Soviets with their large rocket engines, which gives them many months of lead-time, and recognizing the likelihood that they will exploit this lead for some time to come in still more impressive successes, we nevertheless are required to make new efforts on our own. For while we cannot guarantee that we

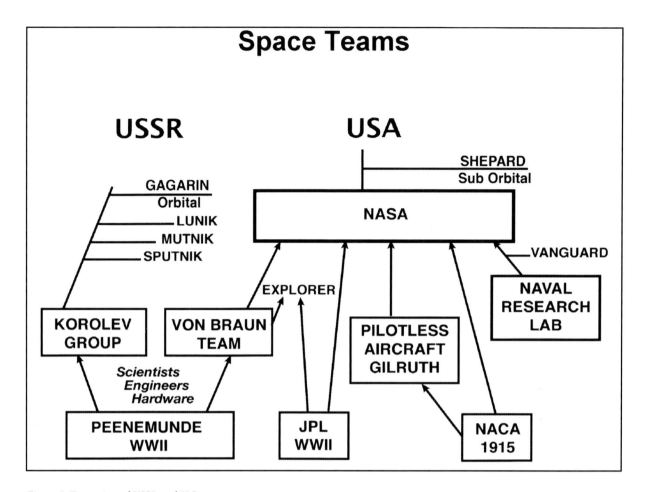

Figure 5. Formation of USSR and U.S. space teams.

shall one day be first, we can guarantee that any failure to make this effort will make us last. We take an additional risk by making it in full view of the world, but as shown by the feat of astronaut Shepard, this very risk enhances our stature when we are successful. But this is not merely a race. Space is open to us now; and our eagerness to share its meaning is not governed by the efforts of others. We go into space because whatever mankind must undertake, free men must fully share.

First, I believe that this nation should commit itself to achieving the goal, before this decade is out, of landing a man on the Moon and returning him safely to earth. No single space project in this period will be more impressive to mankind, or more important for the long-range exploration of space, and none will be so difficult or expensive to accomplish. Now this is a choice which this country must make,

and I am confident that under the leadership of the Space Committees of the Congress, and the Appropriating Committees, that you will consider the matter carefully.

It is a most important decision that we make as a nation. But all of you have lived through the last four years and have seen the significance of space and the adventures in space, and no one can predict with certainty what the ultimate meaning will be of mastery of space.

I believe we should go to the Moon. But I think every citizen of this country as well as the Members of Congress should consider the matter carefully in making their judgment, to which we have given attention over many weeks and months, because it is a heavy burden, and there is no sense in agreeing or desiring that the United States take an affirmative position in outer space, unless we are prepared to do the work and bear the burden to make it successful.[13]

At the time Kennedy was delivering his address to Congress, Mr. Webb and I were meeting with Joe Charyk, Under Secretary of the Air Force. NASA was about to assume a tremendous responsibility, but the orbiting of John Glenn was still to be accomplished. And the Air Force was questioning our use of the Atlas booster. General Bernard Schriever, who had successfully directed the development of ICBMs, was concerned about whether the thin-skinned (0.010-inch) Atlas (see figure 6) could support the Mercury capsule—if it failed, would our nuclear deterrent remain credible? Of course, if the Atlas failed or if a decision was made not to use the Atlas, John Glenn would not achieve orbit and there would be no U.S. manned flight until another launch vehicle became available. I remember attending a detailed briefing on the structural integrity of the Atlas nose section with and without strengthening. The analysis convinced me that it was safe to proceed with the mission if a bellyband was sweated around the nose cone. Joe Charyk concurred. Ultimately, four manned Mercury capsules were successfully launched by the strengthened launch vehicle. But while we were still in the throes of Mercury, we had to start facing the many daunting challenges of President Kennedy's new mandate.

Aiming at the Moon

Now that the President had recommended a major national effort to land man on the Moon within the decade, major decisions had to be made in a short period of time:

1. How was the mission to be managed?

2. How much of the effort would be performed by NASA? By other government agencies? By industry? By universities and other nonprofits?

3. What were the long poles in the tent? That is, what projects required immediate attention?

4. How were we to resolve a large number of technical issues?

One of the keys to the success of this daunting program was NASA's internal management. NASA had four program offices when Keith Glennan was Administrator. They were Advanced Research Technology (headed by Ira Abbott), Life Sciences (Clark T. Brandt), Launch Vehicle (Don R. Ostrander), and Space Flight Programs (Abe Silverstein). Each of the research and flight centers reported to one of these program directors (see figure 19).

Each program office had its own budgeting and cost controlling, as well as its own research centers. It was decided to shift the entire project and program responsibility for NASA to the Associate Administrator. The shift of personnel from Silverstein to me followed. For the next seven years, project approval documents spelling out objectives, costs, and schedules were issued by this office and signed by the Associate Administrator (me) for all NASA activities. A fuller account of NASA management during this period is given in chapter 5. Second, much of NASA's effort required close coordination with the DOD. A NASA-DOD board had been established in the Eisenhower administration with Hugh Dryden and Harold Brown as cochairmen. The board was called the Aeronautics and Astronautics Coordinating Board, or AACB. In the spring of 1961, I became the NASA cochair and Rubel the DOD one; both of us were closer to day-to-day management issues than our predecessors.

The most critical decision was the appointment of the Apollo manager. Discussions were held with the Air Force regarding Bernard "Bennie" Schriever and with the Navy regarding Levering Smith, who directed the Polaris and Poseidon submarine-launched ballistic missile programs. Levering was disappointed that he was still a captain. We succeeded in getting Levering promoted from captain to rear admiral, but not in acquiring him as a NASA manager.

While Dr. Dryden was away, Jim Webb and I had a conversation with Wernher von Braun about the possibility of his directing the lunar landing program. When Dryden returned, Webb asked me to try the idea on him for size, and his answer was,

---

13. "Special Message to Congress on Urgent National Needs," 25 May 1961, *Public Papers of the Presidents of the United States, John F. Kennedy, January 20–December 31, 1961* (Washington, DC: U.S. Government Printing Office, 1962).

Figure 6. Launch of Friendship 7 *on 20 February 1962 for the first American manned orbital spaceflight. John Glenn was on his way to becoming the first U.S. astronaut to orbit Earth. (NASA Image Number 62PC-0011)*

"You and Jim can do what you want, but I'll retire if he's given the job." I suggested Brainerd Holmes. I had known him at Radio Corporation of America (RCA), where he had been in charge of the Ballistic Missile Early Warning System (BMEWS). It was a complex, high-technology project with large-scale construction in Scotland, northern Greenland, and Fairbanks, Alaska. I had a conversation with Brainerd's boss, Art Malcarney, Executive Vice President for Defense Affairs, and he reluctantly helped us arrange a meeting with Brainerd at the Metropolitan Club. Jim Webb and I attended, and thanks to Webb's great salesmanship, Brainerd accepted the position a week later and took the reins in October.

Extensive Planning

The period between the President's recommendations to Congress in May 1961 and the arrival of Brainerd Holmes in October 1961 involved extensive planning as NASA initiated its greatly expanded program. Three of the efforts were carried out by Center-Headquarters committees established by the Associate Administrator; one was a product of Langley Research Center, and one was conducted jointly with the Department of Defense. These committees and their studies are listed below:

- "Various Vehicle Systems for the Manned Lunar Landing Mission," completed 10 June 1961. A study initiated on 21 May 1961 and chaired by Bruce Lundin.

- "A Feasible Approach for an Early Manned Lunar Landing," completed 16 June 1961. A study initiated on 2 May 1961 and chaired by William Fleming.

- "Earth Orbital Rendezvous for an Early Manned Lunar Landing," completed August 1961. A study initiated on 20 June 1961 and chaired by Donald Heaton.

- "Manned Lunar Landing Through Use of Lunar Orbit Rendezvous," completed 31 October 1961. A Langley Research Center report by John Houbolt.

- "Large Launch Vehicles Including Rendezvous," completed 24 September 1962. A joint DOD-NASA study initiated on 23 June 1961 and chaired by Nicholas Golovin (NASA) and Lawrence Kavanau (DOD).

During May, June, and July, when the first two studies (Lundin and Fleming) were under way, there were three Saturn launch vehicles under consideration. The two-stage Saturn I having eight H-1 engines in its first stage and six RL-10 engines in its second stage was of use only for Earth-orbiting payloads.

The advanced Saturn had two configurations: the C-2, for which NASA had contracted, and the C-3, a more powerful configuration. Both versions were in design and had similar first and third stages. The first stage in each used two F-1 engines, and the third stage in each was similar to the Saturn I second stage. However, the second stage of the C-2 used two J-2 hydrogen-oxygen engines with a total thrust of 400,000 pounds. The second stage of the C-3, with four J-2 engines, had a total thrust of 800,000 pounds.

"Various Vehicle Systems for the Manned Lunar Landing Mission," a Study Chaired by Bruce Lundin, 10 June 1961

The report of this committee first discusses the use of the launch vehicle, at that time undergoing design, and the use of rendezvous in both Earth and lunar orbit. Then there is an outline of the pros and cons of the following options:

I.   Earth rendezvous with Saturn C-2s

II. Earth rendezvous with Saturn C-3s

III. Lunar rendezvous with Saturn C-3s

The report states in conclusion that the committee strongly recommends the second alternative. Excerpts from the report follow:

In response to the request of the Associate Administrator on May 25, 1961, a study has been undertaken to assess a wide variety of systems for accomplishing a manned lunar landing in the 1967–70 time period. This study has, as directed, placed primary emphasis on the launch vehicle portions of the [systems, including] vehicle sizes, types and staging. In addition a number of variations on the use of rendezvous to add flexibility and improve energy management in the lunar mission have been considered. The results of this study are the subject of this report.

Mission staging by rendezvous has been the subject of much investigation at Marshall,

Langley, Ames, Lewis, and JPL. The work has concerned itself with analytical and simulator studies of orbital mechanics, and control and guidance problems as applied to rendezvous. Such critical questions as launch timing, and automatic and piloted guidance of the vehicles to a rendezvous have been carefully analyzed. Orbital refueling as well as attachment of self-contained modules have been considered.

Because the use of rendezvous permits the accomplishment of a given mission in a number of different ways employing different launch vehicles, the various groups working on rendezvous have arrived at a number of different concepts for accomplishing the lunar landing mission. The assumptions made by the different groups with regard to such parameters as return weight, specific impulse, etc. were however, consistent to the extent that meaningful comparisons can be made between the different concepts.

The vehicles considered were restricted to those employing engines presently under development. These vehicles are:

a. Saturn C-2 which has the capability of placing about 45,000 pounds in earth orbit and 15,000 pounds in an escape trajectory;

b. Saturn C-3 which has the capability of placing about 110,000 pounds in earth orbit and 35,000 pounds in an escape trajectory.

Lunar [Orbit] Rendezvous

A concept in which a rendezvous is made in lunar orbit possesses basic advantages in terms of energy management and thus launch vehicle requirements. This approach involves placing the complete spacecraft in orbit about the moon at a relatively low altitude. One or two of the three-man crew then descends to the lunar surface; after landing the capsule performs a rendezvous with that portion of the spacecraft which remained in lunar orbit. The lunar capsule is, of course, left behind on the return trip of the spacecraft to earth.

The basic advantage of the system is that the propellant required for the lunar landing and take-off is reduced which in turn translates into a reduction in the amount of weight which must be put into a lunar escape trajectory. The escape weight saving achieved is related to the fraction of the spacecraft weight which is retained in lunar orbit. The actual weight saving which can be realistically achieved by this method can only be determined after detailed consideration of the design and integration of the complete spacecraft. Calculations suggest, however, that the amount of weight which must be put into an escape trajectory for a given reentry vehicle weight might be reduced by a factor of two by use of the lunar rendezvous technique. The earth booster requirement might therefore be reduced to one C-3 with lunar rendezvous or two to three C-3's with earth rendezvous. [I had already received a letter advocating this approach from John Houbolt dated 19 May 1961.]

Advantages and Disadvantages Peculiar to Methods Considered

I. Earth Rendezvous with C-2's (5–7 vehicles required)

a. *Advantages*

1. Fast reliability build up due to higher firing rate

2. Assured launch capability from shore bases

b. *Disadvantages*

1. Large number of vehicles required

2. Long time maintenance in orbit and long exposure to space hazards (up to six months with present AMR, Atlantic Mission Range, pad planning.

II. Earth Rendezvous with C-3's (2–3 vehicles required)

a. *Advantages*

1. Only 1 or 2 rendezvous operations required—simpler, less maintenance,

and exposure time compared to C-2 vehicles systems.

2. Vehicle has single shot lunar orbit mission capability.

3. Could possibly launch from AMR

b. *Disadvantages*

1. Requires a new second stage compared to the C-2.

III. Lunar rendezvous with C-3 (1 vehicle required)

a. *Advantages*

1. Energy, and thus vehicle size, potentially reducible by the order of 50%.

2. Direct monitoring of landing operation possible from orbiter. (Wide band communication available to enhance monitoring).

b. *Disadvantages*

1. A non fail-safe rendezvous

2. Does not have effective assistance from surface tracking and communication networks for the rendezvous maneuver.

3. No growth potential for increased mission requirements.

Of the various orbital operations considered, the use of rendezvous in earth orbit by two or three Saturn C-3 vehicles (depending on estimated payload requirements) was strongly favored. This preference stemmed largely from the small number of orbital operations required and the fact that the C-3 is considered an efficient vehicle of large and future growth.[14]

It's interesting to note that as early as 10 June 1961, a Headquarters-Center study group made such a strong representation for Lunar Orbit

Rendezvous but then rejected the mode out of hand because there could be no backup in case of failure to rendezvous. There could be other single-point failures, such as a propulsion explosion when lifting off the lunar surface. It would take another year for this mode to become accepted in NASA and still more months before the White House allowed NASA to proceed. John Houbolt's concept took a long time aborning.

"A Feasible Approach for an Early Manned Lunar Landing," a Study Chaired by Bill Fleming, 16 June 1961

The study was to be accomplished as rapidly as possible and in no more than four weeks. Excerpts from the terms of reference follow:

There is hereby established an Ad Hoc Task Group that has the immediate responsibility for determining for NASA in detail a feasible and complete approach to the accomplishment of an early manned lunar landing mission. This study should result in the following information:

1. Identification of all tasks associated with the mission.

2. Identification of the interdependent time phasing of the tasks.

3. Identification of areas requiring considerable technological advancements from the present state-of-the-art.

4. Identification of task for which multiple approach solutions are advisable to insure accomplishment.

5. Identification of important action and decision points in the mission plan.

6. Provision of a refined estimate by task and by fiscal year of the dollar resources required for the mission.

7. Provision of refined estimates of in-house manpower requirements by task and by fiscal year.

---

14. Committee chaired by Bruce Lundin, "Various Vehicle Systems for the Manned Lunar Landing Mission," report to Robert C. Seamans, Jr., NASA Associate Administrator, 10 June 1961.

8. Establishment of tentative in-house and contractor task assignments accompanying the dollar and manpower resources requirements.

The following gross programmatic guidelines shall serve as a starting base for the study:

1. Manned lunar landing target date—1967—determine if feasible.

2. Intermediate missions of multi-manned orbital satellites and manned circumlunar missions are desirable at the earliest possible time.

3. The nature of man's mission on the moon as it affects the study shall be determined by the Task Group, i.e., the time he is to spend on the moon's surface and the tasks that he shall perform while there.

4. In establishing the mission plan, evaluate use of the Saturn C-2 as compared to an alternate launch vehicle having a higher thrust first stage and C-2 upper stage components.

5. The mission plan should include parallel development of liquid and solid propulsion leading to a Nova Vehicle and should indicate when the decision should be made on the final Nova configuration.

6. Nuclear powered launch vehicles shall not be considered for use in the first manned lunar landing mission.

7. The flight test program should be laid out with adequate launchings to meet the needs of the program considering the reliabilities involved.

8. Alternate approaches should be provided in critical areas.[15]

Bill Fleming submitted his 510-page, comprehensive report entitled "A Feasible Approach for an Early Manned Lunar Landing" the week

following the submission of the Lundin study. The report did not attempt to find the optimum configuration; rather, it attempted to include all facets of the lunar landing missions, such as spacecraft, launch vehicles, ground support, life and space sciences, and the recruiting and training of astronauts.

For the purpose of this study, a direct flight to the lunar surface using a Nova launch vehicle was assumed. Intermediate-size vehicles were also assumed within the configuration 1 or 2 (C-1 or C-2) category. Configurations 1 and 2 were sized for orbital and circumlunar flights, respectively.

The Sequenced Milestone System, SMS, was used to determine critical areas from a timing and reliability standpoint and to obtain budgetary estimates including the overall total cost. The categories established in the study were the development, fabrication, and testing of all flight hardware; the facilities required for testing and launching the vehicles; the selection and training of the astronauts; the conduct of satellite missions for obtaining necessary environmental data for the lunar mission, especially on the level of radiation en route to the Moon; and the surface conditions on the Moon.

Twelve hundred tasks were specified, and the timing, manpower, and cost were estimated for each. It was determined that land acquisition and facility construction were the "long poles in the tent." The report noted that it was essential to determine the location of all major facilities as soon as possible and to conduct land acquisition, architectural designs, and construction as rapidly as possible.

During the first six months, according to the study, NASA had to accomplish the following:

a. Assign program management and system responsibility.

b. Obtain reentry heating data for the design of Apollo.

c. Get the contract for Apollo and the C-3 first and second stages.

d. Establish flight crew make-up, selection techniques, and training plan.

---

15. Bill Fleming, "A Feasible Approach for an Early Manned Lunar Landing," 16 June 1961.

e. Accelerate the F-1 engine funding.

f. Initiate construction of a wide variety of facilities. These include a new center for spacecraft development and astronaut training, a launch facility with a vertical assembly building, and antennae for ground tracking and communication. Construction had to be hit hard and soon.

The study was based on having facilities far enough off shore to minimize noise and provide safety for those on shore; it also included a vertical assembly building with launching pads over a mile away (see figures 7 and 8). Consideration was given to noise levels in inhabited areas for both Cumberland and Merritt Islands (see figures 9 and 10).

The Fleming report listed three caveats for accomplishing this mission within a six-year period: immediate funding, no major catastrophes, and relief from labor slowdowns.

The study concluded that a manned lunar landing was feasible in the 1967 time period but that major management decisions and actions were required in the first six months. The total cost was estimated to be $12 billion. Critical data were needed on the amount of solar-radiation protection required for the astronauts and on the lunar surface's characteristics.

Mercury Moves Ahead

During the planning and buildup for Apollo, NASA, and particularly the Space Task Group, had to keep focused on all the details of the Mercury Program. Three flights remained in 1961, one of which was manned. Gus Grissom was scheduled for a Mercury Redstone in July. This mission was followed by an unmanned single-orbit test of the Mercury Atlas in September, and a three-orbit mission was scheduled in October with the chimpanzee Enos in the driver's seat.

Figure 7. An offshore launch facility, from the Fleming study.

Friday, 21 July 1961, MR-4, Virgil T. "Gus" Grissom

Gus Grissom and his backup, John Glenn, along with Shepard, had undergone refresher centrifuge training in April, so they were all set for the g forces to be experienced during liftoff and reentry on their next Mercury Redstone missions. Gus and John went back to work right after Alan's flight. The astronauts exercised themselves and the Mercury systems in the simulated high-altitude chamber. Medical data were obtained as they checked the communications, practiced using the manual controls, and simulated complete missions. Each astronaut completed over 100 simulated flights before Gus's flight on 21 July.

Egress from the capsule had required the removal of a bulkhead, followed by a climb through the antenna compartment—difficult for a healthy astronaut, but precarious for an injured one. For this reason, a side hatch was developed with 70 bolts, each with a detonating fuse. When a pin was removed in the cockpit, a fist force of 5 or 6 pounds would open the hatch. In addition, the two 10-inch side windows were replaced by a single trapezoidal window, giving the astronauts nearly 30 degrees of forward vision—up, down, and sideways. Originally, Mercury was going to have a periscope, but no windows; however, the astronauts rebelled at being "Spam in a can." Now they truly could be Earth and sky observers.

Shepard's flight had been overloaded with tests of manual control. Grissom's 10 weightless minutes were to be spent with as much visual observation as possible. There were weather holds on the 18th and 19th, and even on the 21st conditions weren't ideal, but liftoff occurred at 7:20 a.m. The flights went according to plan until *Liberty Bell 7* was afloat following reentry. How it happened is still the subject of speculation, but the hatch blew off as the rescue helicopter approached. The capsule started taking on water as Gus attempted to fasten the helicopter cable. The capsule became too heavy for the helicopter to lift, and Gus started to submerge. On the third try, he was barely able to grab the collar and be pulled to safety. The valve on his suit had not been turned off, so it had filled with water, but Gus was okay. *Liberty Bell 7* lay on the ocean floor for

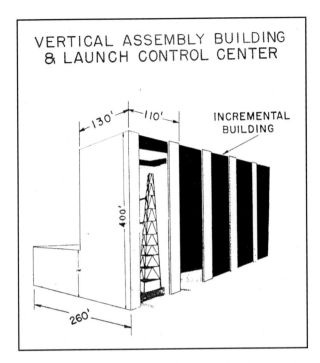

Figure 8. A Vertical Assembly Building, from the Fleming study.

nearly 40 years until it was rescued by entrepreneurs who put it on display.[16]

Korolev Scores Again

Two weeks after Gus Grissom's suborbital flight, Soviet cosmonaut Gherman S. Titov became the first space explorer to stay in orbit over 24 hours. The flight of Vostok II, four months after Gagarin's famous first endeavor, showed us that the Soviets were in earnest and moving toward major accomplishments in space. Korolev was the mastermind of a progressive program that was pressing ahead on all fronts. At this early stage, we didn't know his name or background, but we knew that the Soviet space program was managed skillfully and with imagination.

Wednesday, 13 September 1961, One Orbit Unmanned, MA-4

The Mercury capsule was launched by an Atlas booster, hence the mission was designated "MA". The first launching took place on 13 September

---

16. Swenson, Grimwood, and Alexander, *This New Ocean*, p. 367.

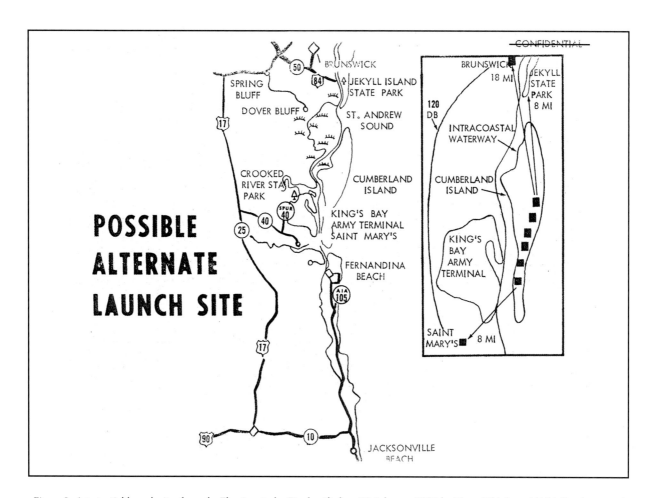

Figure 9. A potential launch site, from the Fleming study. (Declassified on 28 February 2005 by Norm Weinberg, NASA Headquarters.)

1961. The so-called "thin skin" Atlas had a modified nose section to better carry the capsule weight. The flight proceeded through maximum dynamic pressure after 52 seconds of flight. All systems were go, and a peak velocity of 17,600 mph was reached. The maximum acceleration was 7.6 g's. The orbit was slightly lower than planned, but acceptable, so the flight continued. During the flight, simulated crewmen placed on board the craft continued to "breathe" oxygen and produce moisture and carbon dioxide. High oxygen usage was reported early, and the tracking station in Zanzibar reported that only 30 percent of the primary supply was left. The retrorockets were fired in the vicinity of Hawaii, the drogue and main parachutes opened at the appropriate altitudes, and the destroyer *Decatur* made the recovery. The cause for the excessive use of oxygen was discovered. Vibration caused a flow-rate

handle to become dislodged from detent. A new emergency-rate handle with positive latching was devised for later missions. The mission was judged a complete success.[17]

Saturday, 7 October 1961, MA-5, Three-Orbit Chimpanzee Mission

Some questioned the need for another test mission prior to manned orbital flight. By this time, the Soviets had achieved their second manned orbital success with cosmonaut Titov. Wouldn't the United States look ridiculous with still another chimp at the controls? The Space Task Group team, headed by Bob Gilruth, was adamant: we had to stick to our plan and not be rushed. There was a fairly long list of modifications required as a result of MA-4,

---

17. Swenson, Grimwood, and Alexander, *This New Ocean*, p. 398.

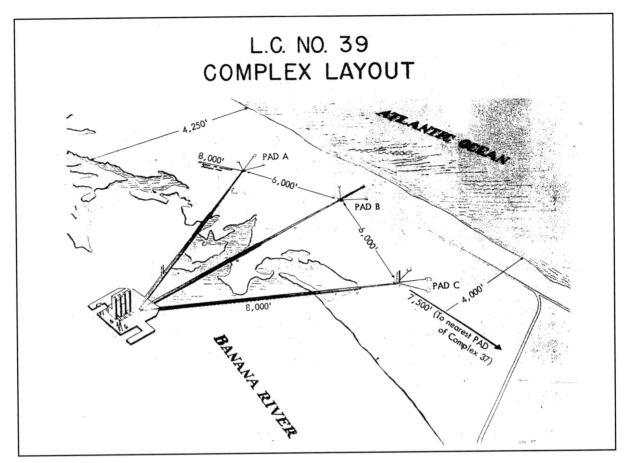

Figure 10. Layout of Launch Complex 39, from the Fleming study.

and the new trapezoidal window had not been tested at orbital speeds. Enos the chimpanzee would arrive with his own metal-plastic pressure coach that was connected to the suit circuit of the regular environmental control system.

The mission plans approximated as nearly as possible those of the upcoming first manned flight. Orbital insertion took place at an altitude of 100 miles, 480 miles from the Cape. The capsule would travel around the world at 17,000 miles per hour and, after 4 hours and 32 minutes, would fire its retrorockets over the Pacific. On reentry, the outside temperatures would reach 1,260° F on the capsule's section, 2,000°F on the antenna housing, and 3,000°F on the heatshield. Enos and his chariot performed according to plan until a yaw reaction jet malfunction, and the flight was terminated successfully after the second orbit.[18]

By October's End–In Progress or Completed

Two of the major Apollo studies were complete. The study teams were composed of Headquarters and field personnel and were needed to establish goals and priorities during the interregnum before the Headquarters program directors were on hand and a new organizational structure could be erected. Other complete actions included the following:

1. Initiation of contractor design studies for Mercury II (later Gemini). This spacecraft would be launched by Titan II and carry two astronauts.

2. Completion in September of studies to determine the Apollo launch site. The NASA-DOD team was chaired by

---

18. Ibid., p. 398.

General L. I. Davis, commander of the Atlantic Missile Range, and Dr. Kurt Debus, director of NASA activities at Cape Canaveral. Six sites were reviewed. Cumberland Island, Georgia, and Merritt Island, across the Indian River from Cape Canaveral, were the serious contenders. The ad hoc group recommended Merritt Island because of its proximity to the Air Force facilities at Cape Canaveral.

3. Initiation of a NASA-DOD large launch vehicle study. In order to conduct large launch vehicle developments of maximum benefit to both NASA and DOD, a comprehensive study was initiated on 23 June 1961. This study was co-chaired by Dr. Lawrence Kavanau from DOD and Dr. Nicholas Golovin of NASA. The ad hoc group examined solid and liquid propulsion, launch vehicles that ranged in size from the Titan series to Nova-type monsters. The group also examined rendezvous options.

4. Completion in July of further rendezvous studies for manned lunar landing. These studies were conducted by the ad hoc group chaired by Colonel Donald Heaton. The report confirmed that by using rendezvous in Earth orbit, the United States could achieve lunar landing at least one year earlier than by a direct ascent to the Moon using a Nova vehicle. However, as many as three launches of the Saturn C-3 might be required, as opposed to only two with still another version of the Saturn, the C-4. The C-4 had four F-1 engines in the first stage rather than two. The report also contained a list of major management decisions and actions required during the first six months of the program.

5. Launch of the first stage of Saturn I on 27 October 1961. The first Saturn SA-1 was static-tested at Huntsville, Alabama, in May 1961 and then shipped to the launch site at Cape Canaveral. The 162-foot carrier weighed nearly one million pounds. Its eight H-1 engines lifted its payload of sand on 27 October 1961 and traveled 200 miles downrange. More than 500 different measurements were recorded, and the flight was deemed flawless.

6. Authorization by Congress of 425 excepted positions, raised from 290. Excepted positions didn't come under the aegis of the civil service. Individuals in these positions were hired and fired at the pleasure of the Administration.[19]

It was obvious from the start that NASA's Apollo Program would require a substantial increase in manpower, but as a matter of policy, the major increase should come from the support of other governmental agencies, industry, and universities. Industrial teams would be selected by procurement procedures, which, although somewhat standardized, would be refined for the purposes of NASA's programs. Specifically, there would not be a source *selection* team, but a source *evaluation* team—the Triad of Webb, Dryden, and Seamans would make the final decision (see figure 2). A variety of incentive arrangements were tested as the program evolved and expanded. University participation would normally result from grants, but there were exceptions—for example, in the development of Apollo guidance and navigation, the Massachusetts Institute of Technology operated on a cost-plus contract. One key area had to be the direct responsibility of the government, namely, land acquisition and construction of facilities. NASA had minimal internal capability, and time didn't permit the evolution of such a capability. One day, Mr. Webb came into my office unannounced and ready for travel. He wanted me to join him on an important mission. As we settled into his black limousine (actually a black Checker cab), he explained that we were headed to the office of Lieutenant General William F. Cassidy, commander of the Corps of Engineers. The Army Corps of Engineers builds all manner of dams, waterways, and buildings, and we were hoping to

19. Arnold S. Levine, *Managing NASA in the Apollo Era* (Washington, DC: NASA SP-4102, 1982), pp. 318–320.

enlist their support for land acquisition and construction of facilities. As I remember, the meeting was relatively short; we explained our mission and its needs, and General Cassidy assured us that the Corps could satisfy our requirements and wouldn't require approval for extra billets (manpower openings). Their performance was truly remarkable.

Manned Spacecraft Center

Each of the three lunar landing studies emphasized the requirement for early construction of facilities if the 1967 date was to be achieved. And, of course, the construction couldn't commence without a decision on those facilities' locations and requirements. The location of the launch facilities on Merritt Island had already been discussed. Also mentioned was the need for a manned spaceflight center to match the launch vehicle establishment under Wernher von Braun in Huntsville, Alabama. The Space Task Group was managing the Mercury Project and would serve as the nucleus for all manned spacecraft development, astronaut training, and space operations. NASA's Langley Research Center had spawned the Space Task Group, but growth in the Tidewater region of Virginia was limited by several factors, such as the lack of available land, local personnel, technology base, and university support. NASA needs and political benefits led to Houston, Texas. The districts of Tiger Teague, chairman of NASA's authorization committee, and Albert Thomas, chairman of our appropriation subcommittee, shared Houston and its environs. In addition, the land for the new center was donated to NASA as an additional come-on. Rice University would be close by, an important part of the total package.

Shipment of Launch Vehicles and Spacecraft

One of the important issues facing NASA was the means for shipping the large, heavy, and somewhat delicate stages of the boosters and spacecraft modules. The Marshall Center examined the feasibility of dirigibles lashed together with the space hardware hanging in between. They even looked into the possibility of acquiring Lakehurst, New Jersey, where dirigibles used to land before World War II. Another avenue Marshall investigated was the development of special aircraft. When these possibilities floated to Washington, they became nonstarters. Water became the way to go, but what type of vessels should be used? Roll-on, roll-off-type barges had many advantages and were selected. With relatively low draft, vessels from Marshall could reach the Gulf of Mexico by traveling on sections of the Tennessee, Ohio, and Mississippi Rivers. Similarly, Houston, with a waterway to Galveston, is nearly on the Gulf; once there, cargo can readily be shipped to Cape Canaveral.

Our next-door neighbor, Marvin Coles, was the chief lobbyist for the Maritime Industries of America. I came home one evening after testifying on issues of transportation to find him practically on my doorstep. To paraphrase, he said, "I hope you're not going to start a U.S. trip to the Moon by using foreign shipping." My rejoinder was, "How about helping our country find suitable shipping rather than getting in a swivet over what we might do?" He did, and our needs were satisfied. The second stage of Saturn I and the nearly identical third stage of Saturn V were manufactured by McDonnell Douglas in Santa Monica, California. Three stages could be shipped by sea, as was necessary for the second stage of the Saturn V manufactured by North American Aviation (NAA), also in California. However, an enterprising small company, Aero Spacelines, perhaps hearing of Huntsville's efforts, modified a Boeing 377 Stratocruiser into the most unlikely vehicle imaginable. The cargo area of the fuselage was doubled in volume, giving the plane its informal name, "the pregnant guppy." NASA was all for the use of the plane if it could be certified by the Federal Aviation Administration. One day, I was asked to approve a voucher for a small amount. The company had run out of assets and, for lack of fuel, couldn't complete the required testing. NASA approved the funds, the license was obtained, and the "pregnant guppy" provided years of service.

First-Stage Construction Site

About that time, we selected Houston as the location for the Manned Spacecraft Center. I received a call from Wernher von Braun about a 45-acre building in the outskirts of New Orleans. The Michoud Plant was on property fronting the Mississippi River. The building had been used by Higgins for shipbuilding during World War II and by Chrysler for making tank engines during the Korean War, and it was currently idle. After further investigation by the Marshall team, along with Chrysler and Boeing, the contractors for the first stages of the Saturn I and the Saturn V, respectively, all agreed that there was room for the fabrication of both booster stages. However, there was a need for additions to the structure. The Corps of Engineers built a high bay area for static testing and constructed a partition separating the contractor manufacturing areas with one portal between them called Checkpoint Charlie. Both the engineering and manufacturing space required extensive rehabilitation. The washrooms were not only segregated between blacks and whites, but I also found, on an inspection tour, that the doorways were significantly narrower for blacks. Needless to say, all facilities were henceforth integrated.

Mississippi Test Facility

Just across the Pearl River from Louisiana was ideal property for static testing of all Saturn I and Saturn V stages, as well as test stands for the F-1 and J-2 engines. This large tract of land in Mississippi had about 600 inhabitants who had lived there for generations. When the Corps started acquiring the land by eminent domain, questions arose as to where the inhabitants could move and still trap muskrats, their primary livelihood. Throughout its properties, NASA attempted to preserve the native habitat. On Merritt Island, where the government acquired 55,000 acres, many of the orange groves were maintained and leased back to the former owners. Before every launching, a special whistle warned the birdlife of the upcoming earsplitting noise. Many eagles' nests in the area are still active. Wherever possible, NASA maintained the land it acquired for both human use and natural habitat.

Kennedy's Spaceflight Center

However, the largest structure of all was the Vertical Assembly Building (VAB), which was to be constructed on Merritt Island, close to Cape Canaveral. The building could house four Saturn V vehicles at any one time, and each could exit the building by a separate set of doors. The Saturn V stood 360 feet tall, and when it was mounted on its transportation, along with the umbilical tower, well over 400 feet of height was required. The central section of the building was 525 feet tall and covered 8 acres. It was mounted on 4,225 piles, each driven to a depth of 150 to 190 feet. A new type of vibrating pile driver was used facilitating penetration into sandy soil.[20] Three architecture teams were used for the total facility, which, in addition to the main section, had a low bay area and the launch control centers.

Although the VAB was the largest building by volume in the world, there was nothing nearby to make it appear so. Also, when you stood on the roof, it didn't appear high because the roof lines extended so far that they blended with the land and sea out to the natural horizon. But take the elevator to the 52nd floor and walk across the catwalk just under the roof, and vertigo could suddenly take hold. Visitors always remembered their trip to the 52nd floor and their view downward of large rocket structures and diminutive people.

Important program milestones had recently been achieved. Many more were pending. Brainerd Holmes arrived amidst plenty of activity. Office space for him and his staff would be required in the District of Columbia. It was already decided that the Office of Manned Spaceflight (OMSF), of necessity, had to be close to the rest of NASA Headquarters and to Congress. Accommodations for Brainerd's office were found near George Washington University.

The development plan for Mercury II became available for Brainerd's review in November. The primary purpose of Mercury II was to gain experience with orbital maneuvers, including the rendezvous and docking with the unmanned test vehicle, Agena. An appropriate award was offered for an appropriate name for Mercury II. "Gemini," referring to the heavenly twins Castor and Pollux, won the special award of a bottle of Old Fitzgerald hands down.

National Headquarters Reorganization

On 1 November, NASA announced a major Headquarters reorganization with five new program offices. The new offices (see figure 20) were Manned Space Flight, under D. Brainerd Holmes;

---

20. Charles Murray and Catherine Bly Cox, *Apollo: The Race to the Moon,* New York: Simon & Schuster, July 1989), p. 319.

Space Sciences, under Homer Newell; Space Applications, under Morton Stoller; Advanced Research and Technology, under Ira Abbott; and Tracking and Data Acquisition, under Edmond Buckley. These program offices and their field installations reported directly to the Associate Administrator. There was one general manager for all NASA research, development, fabrication, and operations. He was called the Associate Administrator. That was my job from 1 September 1960 to 5 January 1968. With this arrangement, I was able to work directly with Bob Gilruth as he moved the Space Task Group to Houston and with Kurt Debus as he formed a new Center at the Cape. I also spent time with Wernher von Braun at Huntsville as he transformed his Army-type arsenal into a project-type institution. Up until 1963, the Johnson Center was still dotted around Houston. At the Cape, launch and administrative buildings were under construction, and at all three Centers, major contracts were under negotiation. It was a time of flux as sound procedures were being formulated for the difficult operations ahead.

John Houbolt, Spokesman for Lunar Orbit Rendezvous

On 15 November, I received another letter from John Houbolt.[21] I had received his first letter on 19 May 1961, even before Kennedy had delivered his special message to Congress. Houbolt acknowledged that contacting me was "unorthodox" and that I might feel I was "dealing with a crank." As a matter of fact, my first impulse upon receiving his letters was to call Tommy Thompson, Director of the Langley Research Center, and ask him to get John off my back. However, I saw great merit in lunar orbit rendezvous, the mode of operation so strongly proposed by John. In his second letter, after considerable fulminating about the rocket engines and launch vehicles under development, John recommended the following steps:

1. Get a manned rendezvous experiment going with Mercury MK II (soon to be called Gemini).

2. Firm up the engine program suggested in his letter and attachment, converting

the booster to these engines as soon as possible. (John didn't know we were about to approve the C-5.)

3. Establish the concept of using a C-3 and lunar rendezvous to accomplish the manned lunar landing as a firm program. (I was trying my best, but controversial decisions of this type cannot be made by decree. The next step was to bring this message to Brainerd in such a manner that he'd respond positively to the concept of rendezvous in orbit around the Moon.)

John not only wrote me a letter, but, with two others, he also wrote a comprehensive, 97-page report entitled "Manned Lunar Landing Through the Use of Lunar Orbit Rendezvous." It was dated 31 December 1961 and gave a thorough outline of his rendezvous research at Langley Research Center. In addition, John had made a most favorable impression when I visited Langley five days after my swearing-in. He had explained the orbital maneuvers and noted the 50-percent savings in weight. The biggest stumbling block in people's minds was the absolute requirement for a successful rendezvous in lunar orbit. This maneuver seemed quite manageable to me as a result of my experience at the Massachusetts Institute of Technology (MIT) and the Radio Corporation of America (RCA). At MIT, I directed projects on airborne missiles; at RCA, the DOD SAtellite INTerceptor project, SAINT. Obviously there was some risk in this portion of the mission, but the overall trip to the Moon and back was loaded with risk, and its totality had to be minimized.

Obtaining Systems Capability

As can be readily seen from John Houbolt's concerns, the Apollo Program office had an acute need for a systems capability. This need for an ability to conceive and define quantifiably all the elements required to complete the mission was manifested early on, and we attempted to recruit an in-house systems team within the Office of Manned Spaceflight. Fortunately, through the good services of Mervin Kelley, special consultant to Jim Webb, this effort

21. John Houbolt to Robert C. Seamans, 15 November 1961, Robert Channing Seamans, Jr., papers, MC 247, Institute Archives and Special Collections, MIT Libraries, Cambridge, MA.

yielded Joseph Shea (later to become the Apollo chief systems engineer) and a contract laboratory, Bellcom. Merv had been the director of AT&T's Bell Laboratories. He helped open the door to James Fisk, the then-director. I visited the Bell Labs, explained our needs and deficiencies to Jim, and proposed the establishment of a system laboratory in Washington, similar to but smaller than Sandia, their laboratory set up for the Atomic Energy Commission (AEC). Jim was by no means overjoyed at the thought because some of his key personnel would be required to establish the enterprise. He also recognized that Apollo might have a lifespan of little more than a decade. He said that any hires would become permanent members of the Bell Laboratories, not to be let go if Bellcom's contract was terminated. Hence, they would hire only top-grade personnel. I agreed and said that's what was needed, but I also recognized that Bellcom would be slow at the starting gate. When formed and manned, the Bellcom group provided top-grade analytical capability that supported system integration at both Headquarters and the Centers.

After receiving his doctorate, Joe Shea had become a member of the Bell Laboratories; more recently, Joe had been responsible for the intercontinental ballistic missile (ICBM) guidance for the Titan program at AC Spark Plug. After Joe's swearing-in as the Apollo chief systems engineer, he and Brainerd came to my office. He claims my instructions were to "sell lunar orbit rendezvous" to the manned flight organization. This was not an easy assignment because most members of von Braun's Marshall Center and Gilruth's Manned Spacecraft Center (then the Space Task Group) had taken a strong stand for either Earth orbit rendezvous or direct ascent. Joe's quest took him into the inner circles of the flight centers in Houston and Huntsville and, ultimately, the White House, and his efforts were eventually crowned with success. NASA did utilize lunar orbit rendezvous to achieve a manned lunar landing within the decade, and, in my view, that mode of operation was the only road to success. But while this key decision was in the balance, NASA achieved a manned orbital flight when John Glenn climbed aboard *Friendship 7* for the second time on 20 February. He was followed several months later by Scott Carpenter.

Tuesday, 20 February 1962, MA-6, John Glenn's Orbital Mission

Even before 20 February, John Glenn experienced the sometime vagaries of countdowns. On 27 January, the first countdown for MA-6 commenced, and after a series of holds, the flight was scrubbed. John had been lying on his couch for over 4 hours. During this time, his courageous wife Annie had been at home with her family, and members of the press corps outside were salivating while awaiting news of the flight. Was it to be a grand success with a happy wife or a fatal failure with a family in mourning? Either way, the media were there in eager anticipation.

While this act was in play, Vice President Lyndon Johnson, who was also chairman of the Space Council, was in a limousine nearby, wanting to appear on the scene and bolster Annie's confidence. What might have been a touching scenario was not to be. She didn't want counseling. As soon as John Glenn emerged from the gantry tower, he was advised of the stalemate. Would he please call Annie and tell her that she must welcome the Vice President? His reply was direct: "If Annie doesn't want to see the Vice President, she doesn't have to."

On 20 February, the biosensors were installed on John at 5:00 a.m., and he was soon on the way to the launchpad. By 7:00 a.m., the hatch was bolted in place, and he commented that the weather was breaking up. At 8:05, the time was T minus 60 minutes. After a series of small holds, it became T minus 22 minutes at 8:58 a.m. By then, John Glenn and the blockhouse and Control Center crews were joined by about 50,000 "birdwatchers" on local beaches and an estimated 100 million viewers via TV sets. At 9:47, *Friendship 7* was up and away and John's pulse reached 110 beats per minute. Telemetered signals indicated that the Atlas and spacecraft were performing perfectly. At the maximum dynamic pressure, max-q, John reported, "It's a little bumpy around here." Five minutes after liftoff, *Friendship 7* was "through the gates," and, according to Goddard's computers, conditions were good enough for nearly 100 orbits. Glenn found that he could move about and see well, but he quipped, "I'd like a glass capsule." Weightlessness has its advantages: if his attention was drawn to the panel switches, he'd leave his camera suspended near his head until he could return to taking photos. Within range of the Australian tracking station, he reported feeling fine, seeing bright lights (Perth), and being excited about the "shortest day of his life" (108 minutes from sunrise to sunrise).

Then, when he was over Mexico, he notified the station that a yaw reaction jet was giving him trouble. He had to live with the problem for the rest of the mission by shifting to manual control or just allowing the capsule to drift so that he could save his fuel supplies. Later in the mission, a more serious problem appeared to a ground controller: the heatshield had become unlocked and was held in place only by the straps of the retro package. During the remainder of the flight, a debate continued over whether to jettison the retrorockets after their firing and run the risk of disengaging the heatshield or to leave them on, thereby securing the heatshield until g forces kept it in place but running the risk of damaging the heatshield as reentry heat destroyed the rockets. Christopher Kraft and Walter Williams decided to keep the retro pack in place. (Walter Williams was the mission controller on the early Mercury flights. Christopher Kraft assumed this responsibility later in the program.)

The retros were fired as *Friendship 7* approached the California coast. "Boy, it feels like I'm going halfway back to Hawaii," Glenn exclaimed. As the capsule decelerated and the temperature outside the capsule increased, there was a real fireball outside. Was it the heatshield disintegrating? And then, after passing peak g's, the spacecraft started oscillating wildly and the fuel in the damping control became low. However, the drogue opened at 28,000 feet, and all was well. John and *Friendship 7* were retrieved by *Noa*, a destroyer, and later transferred to the carrier *Randolph*. The recovery team described John's condition as hot; sweating profusely; fatigued; lucid, but not loquacious; thirsty, not hungry. John Glenn was safe and sound.[22]

The President, Jim Webb, and other dignitaries welcomed John when he returned to the Cape. My wife Gene and I were happy to be in the large, welcoming crowd, although we were almost overrun by the media, who were ever straining to close in on America's latest hero.

Thursday, 24 May 1962, MA-7, Scott Carpenter

At 3:45 a.m., Scott and his team boarded his land transportation for a slow ride to *Aurora 7*. The countdown was flawless, with three 15-minute holds solely to wait for ground fog to disperse. During the holds, Scott chatted with his wife, Rene, and their four children. Liftoff and orbital entry were completely nominal, leading Chris Kraft to comment that MA-7 was the most successful flight to date. Scott enjoyed his capsule maneuvers, photographing surface objects over Woomera, Australia, and airglow phenomena. On six occasions, he accidentally activated the high-thrust control jets. So by the end of two orbits, his control fuel was down to 40 percent. He was advised to use his fuel sparingly. So on the final orbit, he went through a long period of drifting. He let the capsule slowly roll until it was close to retrofire. The tracking site at Hawaii instructed Scott to start his pre-retrofire countdown. He advised ground control that he was somewhat behind, as he had spent some time testing his hypothesis about the snowflake particles that John Glenn had seen. Then, as he attempted to lock up his 34-degree nose up and 0-degree yaw, he was in trouble. The automatic control failed. He hurriedly went to fly-by-wire and felt that he was in alignment when he heard Alan Shepard's voice from California asking whether he had bypassed the automatic retro-attitude switch. Carpenter quickly acted on this timely reminder. About 3 seconds after Shepard's call of "Mark! Fire one," the first rocket ignited, followed by the second and third. Actually, *Aurora 7* was yawed 25 degrees to the flightpath, causing a 175-mile overshoot of the landing. The retrorockets were fired 3 seconds late and provided excess power, which accounted for another 75 miles over target. After landing, Scott noticed some water on his tape recorder, and his capsule had a continued list. He had at least an hour's wait, so he decided to abandon ship. Not wanting to open the hatch for fear of sinking, he wormed his way upward through the throat of the spacecraft. To keep cool, he left his suit hose attached and struggled with the life raft, survival kit, and kinked hose before getting his head outside. He was able to disembark, deploy his life raft, and enjoy the presence of the sea bass and gulls until aircraft arrived and two frogmen appeared beside him. They later reported that Scott

---

22. Swenson, Grimwood, and Alexander, *This New Ocean*, p. 423.

appeared "smiling, happy, and not at all tired." After 3 hours, Scott was picked up by an HSS-2 helicopter, which first dunked him in the water before putting him aboard the destroyer *Pierce*. He cut a hole in his socks and proceeded to pace around the deck as the water drained and he talked about his flight. He received the traditional congratulatory call from the President. Scott apologized to him for not aiming more precisely.[23]

Saturn V Defined and Approved for Development

On 25 January 1962, NASA established a standard configuration for Saturn and approved its development program, with Marshall Space Flight Center having responsibility for all three stages using the F-1 and J-2 engines. There were many potential roadblocks that might have derailed the development of the Saturn. Examples include the combustion instability of the F-1 engines and the intractable difficulty of moving the Apollo/Saturn from the Vertical Assembly Building to the launchpad.

In June 1962, NASA announced that the Advanced Saturn had shown considerable growth. The three-stage Saturn booster was originally contracted as the C-2 configuration of Saturn, with Boeing developing the first stage with two F-2 engines, North American the second stage with two J-2 engines, and Douglas the third stage with six RL-10 engines. In the final version, the number of first- and second-stage engines had grown from two to five and the third stage had given up the six RL-10 engines that produced 90,000 pounds of thrust for one 200,000-pound-thrust J-2 engine. When lifting off the pad, the five F-1 kerosene-oxygen engines provided 7.5 million pounds of thrust, and the second-stage burn with hydrogen-oxygen J-2 engines drove Saturn toward orbit with a 1-million-pound thrust.[24] When I was reviewing this transformation with Abe Hyatt, Director of Plans and Evaluation, he asked me whether we should rebid the three contracts because such major changes had been made in the specifications. My answer was, "Not on your life."

Assembling Indoors

The concept of assembling the Apollo/Saturn indoors had great appeal. Construction on the launchpad had been fraught with difficulty. During heavy winds or rain, tarpaulins were dropped around the vehicle for protection against sand and water. And of course, when lightning was striking, there was no assurance that the vehicle would be spared. For this reason, as early as the spring of 1961, plans were approved for a Vertical Assembly Building in which the rocket stages and the Apollo spacecraft would be mated and checked out. Before the VAB was built, launch crews performed their checkout in blockhouses of reinforced concrete adjacent to the pad because analog instruments were utilized and their readout had to be near the vehicle. In the new concept, the on-board instruments might be analog or digital, but conversion to digital would be 100 percent. The RCA computers for this purpose were located in the structure supporting the vehicle and moved with the vehicle from the VAB to the launchpad. The same monitors were used in Launch Control whether the vehicle was in assembly or preparing for launch. During the final 2 minutes before liftoff, the checkout became automatic, surveying all 70,000 instruments to be certain all parts of the vehicle and supporting equipment were within tolerance.

Transportation at a Crawl

But how was the vehicle to be transported? Clearly by rail. However, detailed work on transportation design showed that the combined weight of the vehicle and its supporting structure was too heavy for the steel wheels. Flats would develop overnight, even with all the wheels the design could muster. Next, a review with the Navy led to the conclusion that there was no way to float the assembly in a stable fashion. Our plans for a VAB were nearly scrapped when word came of crawlers used to transport huge draglines in the open pit mines of Appalachia. Although skeptical, a team from the Cape visited an installation. When they asked for a demonstration, they were told to climb

23. Swenson, Grimwood, and Alexander, *This New Ocean*, p. 447.

24. Wernher von Braun and Frederick I. Ordway III, *History of Rocketry and Space Travel* (New York: Thomas Y. Crowell Company, 1966, p. 170.

the ladder and come aboard. Getting impatient, they asked when the crawler would start, only to be told that they had been moving for 5 to 10 minutes. What's more, the crawler had automatic leveling precise enough to keep the vehicle vertical within better than a foot at the top of the capsule 380 feet from the base. Figure 11 shows my sons, Joe and Toby, inside one of the four cars. The individual treads weigh several tons. To ensure success, the roadbeds to the pads had to be excavated 6 feet and filled and rolled with appropriate gravel.[25]

Large Launch Vehicles

Later in 1962 (on 24 September, to be exact), Golovin and Kavanau, cochairmen of the NASA-DOD Large Launch Vehicle Planning Group, submitted the group's final report.[26] The results culminated from major efforts by elements of both organizations. The report recommended "a minimum modification version of the Titan II ballistic missile for the Gemini program."[27] Nearly concurrent with this recommendation was a DOD-NASA agreement recognizing Gemini as a NASA project. The agreement spelled out a Gemini Program Planning Committee to be chaired by the Associate Administrator of NASA and the Under Secretary of the Air Force.[28] This relationship was most fortuitous as longitudinal vibrations of Titan (called the POGO effect, after the movement of a pogo stick) gave visibility and prominence to all hands.

The report also recommended that the Saturn I of NASA be continued but that a Titan III launch vehicle be developed in parallel, thereby providing DOD with a vehicle of similar capability to Saturn I, but with a combination of liquid storables and solids, a launch vehicle more rapidly available in times of crisis.

NASA had approved the Saturn V three months prior to this NASA-DOD report. The report did

support NASA's position that "this development should be pursued with the highest priority . . . lunar orbit rendezvous offers the chance of earliest accomplishment of manned lunar landing. It is quite likely that the pacing item for any rendezvous approach is the development of the launch vehicle, hence the high degree of urgency recommended."[29]

The report then went on to recommend the Nova vehicle with twice the capability of the Saturn V: "Since it is by no means certain that the development of rendezvous operations will advance rapidly enough to provide earliest accomplishment of manned lunar landing, it is recommended that the direct ascent capability be developed on a concurrent basis."[30] The report recommended Nova, but lunar orbit rendezvous had been approved by NASA and, tacitly, by the White House three months earlier. Nova, the mother of all vehicles, was hard to kill. NASA believed that rendezvous could be readily achieved, and it was. Nova was no longer actively pursued. Hence, a major part of the Kavanau-Golovin report that related to very large solids and a 1.5-million-pound hydrogen-oxygen engine became irrelevant.

However, there were three other elements in the report that were most significant, two with which NASA readily agreed and one on the controversial list: 1) automatic checkout, 2) redundancy, and 3) number of flights to man-rate. The report stated that automatic checkout and countdown must be advanced for two reasons: first, to reduce the prelaunch time; and second, to enhance reliability. This observation was certainly true and will be discussed more fully later. The second was for redundancy and specifically for engine out capability. The second Saturn V launching achieved orbit even through two of its five J-2 engines flamed out during the boost phase.

However, the report recommended an excessive number of flights for man rating. Note the follow-

---

25. Murray and Cox, Apollo, *The Race to the Moon*, pp. 98–99.

26. Nicholas Golovin and Lawrence Kavanau, "Large Launch Vehicles Including Rendezvous," NASA-DOD report by The Planning Group, 24 September 1962, Robert Channing Seamans, Jr., papers, MC 247, Institute Archives and Special Collections, MIT Libraries, Cambridge, MA.

27. Golovin and Kavanau, chap. III, section 1.

28. W. Fred Boone, *NASA Office of Defense Affairs*, the First Five Years (NASA, 1970), p. 84.

29. Golovin and Kavanau, chap. II, section 3.

30. Ibid., chap. II, section 5.

Figure 11. The author's sons, Toby (right) and Joe inside one of the treads of the massive vehicle transporter (crawler) at Cape Canaveral on the day after the launching of Gemini 3, 23 March 1965.

ing quotation: "From an examination of the results of the calculation of mission success data analyzed by the large launch vehicle group, it was found that it would take two to three years of flight test and about 25 to 60 launchings to man rate a Saturn I or Saturn V using the reliability growth estimate of this study."[31] The cost alone ruled out such a prodigious number of test launch vehicles. Additionally, the tests are meaningless unless the whole system is scrutinized, as there are many interactions between the spacecraft launch vehicle and the ground environment. The resolution of this issue took place in 1963 and finally gave NASA policies, procedures, and schedules that led to the achievement of President Kennedy's goal. However, the management of manned spaceflight changed hands before this final action was taken.

Nuclear Testing Before the Ban, September 1962

I was a sidebar participant in a meeting between the Atomic Energy Commission (AEC) chairman, Glenn Seaborg, and President Kennedy at the White House in September 1962. There was about to be a worldwide moratorium on atmospheric testing of nuclear weapons. The AEC had nine such tests that it felt were essential. The question on numbers was discussed in some detail, and the President finally agreed to five, not nine. The AEC was bounded by a completion date prior to the treaty date, but when could they start? The test could not be run while Mercury 8 (*Sigma 7*) was in orbit; the radiation level would be too high for communications and for Wally Schirra's health. He was scheduled to launch on 3 October. Everyone turned to me and asked, "Can you meet the scheduled date?" My answer was, "I guess we better." The President closed by saying, "NASA has the October 3rd date and Glenn can test five of his toys."

Saturday and Sunday, 11–12 August 1962, Vostoks 3 and 4, Nikolayev and Popovich

By the time Wally Schirra was prepared for launch aboard *Sigma 7*, the Soviets had performed another key maneuver. On 11 August, Andrian Nikolayev had completed nearly four days in orbit aboard Vostok 3, but that wasn't all. Vostok 4 was launched the next day with Andrian's fellow cosmonaut, Pavel Popovich, at the controls. He maneuvered Vostok 4 within 4 miles of his companion.[32] I can remember that *Aviation Week* carried the story soon after it occurred, and there was much speculation as to Soviet intentions. Were they conducting this dual maneuver to gain experience for their own exploration, or were these tests a prelude for Soviet inspection and possible interception of U.S. satellites? The United States had the unmanned project called SAINT, but orbital tests had not been initiated.

Wednesday, 3 October 1962, *Sigma 7*, MA-8, Walter Schirra

When Wally Schirra buckled himself into his couch and smiled as he saw an automobile ignition key hanging from a safety latch, his flight plans were much more modest than the Soviets'. If all went well, he would be approved for six orbits. The launch phase had a few surprises. Early in the flight, there was an unexpected roll, which stopped just short of the abort condition. Then the booster engine cut off several seconds early, and the escape tower soon sped away, spreading a spotty film on his window. The sustainer engine seemed to burn on and on, although in actuality, it cut off only 10 seconds late. By a small margin, Wally sped faster and higher than any other Mercury pilot.

Temperature control in both cabin and suit had proven difficult in previous missions, and now, according to Position 4 on his controls, the cabin was starting to overheat. Wally wasn't too concerned, comparing his condition to that of mowing his lawn in Texas on a summer's day. However, on his own, he started increasing the knob position a half mark at a time and then waiting 10 minutes to observe the result. By the time he reached Position 8, he was starting to feel cool. Ground control suggested that he go back to 3.5, but he selected 7.5, and there the setting stayed, and he felt comfortable.

The astronauts had no difficulty in using the horizon by day and the stars by night for control-

31. Golovin and Kavanau, chap. III, section 3.

32. House Committee on Science and Astronautics, *Astronautical and Aeronautical Events of 1962: Report of the National Aeronautics and Space Administration* (1963).

ling pitch above or below the horizon. The capsule had to be correctly nosed up for a good reentry. Similarly, roll could be readily brought to horizontal before reentry, but zero yaw was more difficult to achieve, as Scott Carpenter had found in *Aurora 7*. Wally experimented, using ground cues, star patterns, and even the Moon. Night was the most difficult. He found that the correct retrofire attitude placed the planet Jupiter on the upper right of his window, the constellation Grus in a bit from the left side, and the star Fomalhart at the top center.

As the flight continued, he relaxed by powering down and coasting along. He chatted with fellow astronauts as he traveled over various ground and ship stations, and then it was time for the checklist in preparation for his return. When he came into range of the Pacific command ship, he found he still had 78 percent of his control fuel left. Shepard asked how he stood on his checklist. Complete except for arming the rocket squibs, Wally responded. Soon thereafter, the three retrorockets fired in sequence, and he started his descent. He termed his return to the atmosphere as "thrilling." He said that Earth's surface really began to brighten, and most surprising, the "bear" he rode felt as stable as an airplane. Nine hours and 54 minutes after launch, he was hoisted aboard the USS *Kearsarge*. He received congratulatory messages from the President, the Vice President, and his wife Josephina. In the colloquial, it truly was a storybook flight.[33]

Winner of the Final Round—Lunar Orbit Rendezvous

Charles Murray and Catherine Cox's *Apollo: The Race to the Moon* engagingly relates Joe Shea's adventure with the two NASA Centers, MSFC and the Manned Spacecraft Center (MSC).[34] Joe visited Huntsville, home of von Braun's Marshall Space Flight Center, and came away with several good additions to his systems engineering team but a negative response to lunar orbit rendezvous (LOR).

In Houston, home of Gilruth's Manned Spacecraft Center, Joe found kindred spirits for LOR but a negative response to helping him with the systems engineering. So he next hired Chance

Vought Corporation to review Houston's weight estimates. Then, one fateful day, after a series of meetings, the Gilruth team spent 6 hours briefing von Braun and his associates. At the end, there was silence, and John Palp, North American's Apollo manager, was quoted as saying, "I'd like to hear what son of a bitch thinks LOR isn't the right thing to do."[35] Actually, there was still another round of meetings, during which the von Braun team presented its recommendations for Earth Orbit Rendezvous in a 6-hour session. At the end of the day, Wernher rose, complimented his own people, and said that the Lunar Orbit Rendezvous offered the highest confidence for successful achievement within the decade. He pointed out that the separation of capsules for lunar landing and reentry into Earth's atmosphere was bound to simplify the development of the system.

But this was not the end of the story. Brainerd Holmes and his team presented their findings to the Triad (Dryden, Webb, and Seamans), who enthusiastically concurred. A press conference was scheduled for two weeks hence. However, NASA was not to say that a decision had been reached because the President's science advisor was adamantly opposed to LOR—in fact, Jerry Wiesner was never convinced that it was the correct method. Arguments took place at the White House and on an extended trip by the President and Vice President to various NASA manned flight centers. At Marshall, in their large vehicle assembly building, the argument blossomed in front of the nearby press, held at bay behind a rope barrier but within earshot. It wasn't until 24 October 1962 that Jim Webb wrote Jerry that NASA was going ahead with LOR. On 7 November 1962, a press conference was held confirming the decision and announcing that the Grumman Aviation Corporation had been chosen to build the Lunar Lander.

Why Spend Billions?

There's been much conjecture about President Kennedy's motivation when he addressed Congress and recommended a lunar landing and safe return within the decade. Was he a true space cadet fantasizing about a lunar mission from Earth? Or was he

---

33. Swenson, Grimwood, and Alexander, *This New Ocean*, p. 472.

34. Murray and Cox, *Apollo: The Race to the Moon*, pp. 124–128, 133–143.

35. Ibid, pp. 113–120.

impressed with the scientific importance of learning more about our universe particularly our own solar system? Some have suggested that he felt the need for a major effort so that the Soviets would agree to negotiate a joint program. My meetings with the President at the White House on 21 November 1962 and during his visit to Cape Canaveral on 16 November 1963 showed me that he had one straightforward goal, and it wasn't any of the above. He wanted the United States to conduct a major, readily discernable mission in space prior to an equivalent Soviet Union achievement. The Soviets, thanks to Khrushchev's opportunism and Korolev's mastery of Soviet technology, had embarrassed the United States again and again with cleverly devised and well-conceived forays into space. Were we, as they claimed, a degenerating civilization and they the wave of the future? Kennedy wanted to prove it wasn't so.

A meeting with President Kennedy at the White House resulted from the interesting recommendation made by Brainerd Holmes for a supplemental appropriation of $440 million for Apollo. Brainerd had first approached me, stating that a supplement of this magnitude would permit a lunar landing one year earlier, namely, in 1966. I couldn't believe him. In 1961, we had increased Eisenhower's budget request for FY 1962 to $1.8 billion (an increase of more than $670 million); the following year, to 3.7 billion.[36] I was certain that Congress would look askance and that even if the additional funds were available, getting them wouldn't land us on the Moon earlier. I said no but agreed to a meeting with Webb and Dryden, who both also said no. But Brainerd had jump-started the Apollo Program, had generated great esprit de corps, and was considered "good copy" by the media. Soon after we turned down Brainerd's recommendations, an article appeared in *Time* magazine describing an upheaval at NASA: Brainerd and Jim Webb were locked in deadly combat, one of them might have to go, and it wasn't necessarily Brainerd. Such articles can quickly lead to White House interest, and this one did.

The Kennedy Library recently released tapes of the 21 November meeting. David Bell, director of OMB, and Jerry Wiesner were on one side of the table, and Brainerd Holmes, Jim Webb, Hugh Dryden, and I were on the other side when the President arrived. He quickly got to the point:

. . . PRESIDENT KENNEDY: So it's your view that with the four hundred forty million . . . you'd probably, . . . your judgment is you won't really save any time, is that correct?

JAMES WEBB: In the lunar landing I doubt very much if we'd save time. You can schedule it, you can go through the PERT system. Bob Seamans will say yes we . . . we will very likely save from four to six months. But from a general overall look at how these big programs run I doubt if we'd save very much time. Now Bob, I think you ought to say your own views because you are the operating head of this operation.

ROBERT SEAMANS: I think I agree with you Jim, that you can schedule six months earlier but you have to understand what these dates really are. These are dates for the internal management of the projects. They have to be dates that people believe are realistic. I mean, you have to have a fighting chance to achieve these dates but they're by no means dates that you can absolutely guarantee at this time, because this is a development program, and you are learning as you go along and if you crank up too much of a crash program and you start running into trouble, it can take more time to un-sort the difficulties than if it is a better paced program.

JAMES WEBB: A better way to state what I was trying to state. I think we can do in an orderly way what we have scheduled. I think the other will provide quite a series of crises.

. . . PRESIDENT KENNEDY: . . . Do you think this program is the top priority of the agency?

JAMES WEBB: No sir I do not. I think it is one of the top priority programs . . . .

PRESIDENT KENNEDY: Jim, I think it is the top priority, I think we ought to have that

<hr />

36. Table 4.4, "Requests, Authorization, Appropriations, Obligations and Disbursements," p. 131, and table 4.13, "Funding NASA's Program FY1962," p. 138, both in *NASA Historical Data Book*.

very clear. Some of these other programs can slip six months, or nine months and nothing strategic is going to happen, it's gonna . . . . But this is important for political reasons, international political reasons. This is, whether we like it or not, in a sense a race. If we get second to the Moon its nice but its like being second any time. So that if we're second by six months, because we didn't give it the kind of priority, then of course that would be very serious. So I think we have to take the view that this is the top priority with us.

JAMES WEBB: But the environment of space is where you are going to operate the Apollo and where you are going to do the landing.

PRESIDENT KENNEDY: The science . . . . Going to the moon is the top priority project now there are a lot of related scientific information and developments that will come from that which are important. But the whole thrust of the agency in my opinion is the lunar program. The rest of it can wait six or nine months . . . .

JAMES WEBB: Why can't it be tied to preeminence in space which are your own . . . .

PRESIDENT KENNEDY: Because, by God, we keep, we've been telling everybody we're preeminent in space for five years and nobody believes it because they [the Soviets] have the booster and the satellite . . . . We're not going to settle the four hundred million this morning. I want to take a look closely at what Dave Bell . . . but I do think we ought to get it, you know really clear that the policy ought to be that this is the top priority program of the agency, and one of the two things, except for defense, the top priority of the United States Government. I think that is the position we ought to take. Now, this may not change anything about that schedule but at least we ought to be clear, otherwise we shouldn't be spending this kind of money because I'm not that interested in space. I think it's good, I think we ought to know about it, we're ready to spend reasonable amounts of money. But we're talking about these fantastic expenditures which wreck our budget and all these other domestic programs and the only justification for it in my

opinion to do it in this time or fashion is because we hope to beat them and demonstrate that starting behind, as we did by a couple of years, by God, we passed them.

JAMES WEBB: I'd like to have more time to talk about that because there is a wide public sentiment coming along this country for preeminence in space.

PRESIDENT KENNEDY: We do have to talk about this. Because I think if this affects in any way our sort of allocation of resources and all the rest, then it is a substantive question and I think we've got to get it clarified. I'd like to have you tell me in a brief . . . you write me a letter, your views. I'm not sure that we're far apart. I think all these programs which contribute to the lunar program, are, come within, or contribute significantly or really in a sense, let's put it this way, are essential, put it that way . . . are essential to the success of the lunar program, are justified. Those that are not essential to the lunar program, that help contribute over a broad spectrum to our preeminence in space, are secondary. That's my feeling.

. . . ROBERT SEAMANS: Could I state my view on this? I believe that we proceeded on Mercury, and we're now proceeding on Gemini and Apollo as the number one program in NASA. It has a DX priority. Nothing else has a DX priority.

JAMES WEBB: And recommended four point seven billion funds for it for 1964!

ROBERT SEAMANS: At the same time, when you say something has top priority, in my view it doesn't mean that you completely emasculate everything else if you run into budget problems on the Apollo and Gemini. Because you could very rapidly completely eliminate your meteorological program, your communications program and so on. If you took that to too great of an extreme . . . .

JAMES WEBB: And the advanced technology on which military power is going to be based.

HUGH DRYDEN: Mr. President, I think this is an issue. Suppose Apollo has an overrun of five hundred million dollars, to reprogram five hundred million dollars for the rest of the space program would just throw the whole thing away. And I think this is the worry in Jim's mind about top priority.

PRESIDENT KENNEDY: Listen, I think in the letter you ought to mention how the other programs which the agency is carrying out tie into the lunar program, and what their connection is, and how essential they are to the target dates we're talking about, and if they are only indirectly related, what their contribution is to the general and specific things in space. Thank you very much.

Kennedy gets up to leave the room.[37]

President Kennedy and Webb obviously had some difficulty in coming to grips with the other's point of view. The President ultimately recognized that much of NASA's scientific investigation was essential to Apollo. But Jim Webb never agreed with the President that "everything we do ought to really be tied into getting onto the moon ahead of the Russians." Jim asked, "Why can't it [our goal] be tied to preeminence in space?" The President replied that scientists may believe that the two are connected, but not the people of the world who know only that the Soviet Union has the biggest rockets. But the President's bottom line was absolutely clear. He stated that Apollo was NASA's top priority, important for international and political reasons. He added that except for defense, it was, along with one other, the top priority of the U.S. government.

Driving back to NASA Headquarters, I was wondering how Jim's views and those of the President could be brought into concert. Hugh interrupted my thoughts when he said he'd like to prepare the first draft. Later, I received Hugh's version, only a page and a half long and quite general in nature. I attempted a different version, considerably more extensive, with separate sections for manned lunar landing, space science, advanced research technology, university participation, and international activity, with extensive summary and conclusions sections.

I came right to the point in the second paragraph of my nine-page letter: "The objective of our national space program is to become pre-eminent in all important aspects of this endeavor and to conduct the program in such a manner that our emerging scientific, technological, and operational competence in space is clearly evident."[38]

I then followed with paragraphs that detailed what we had to do to become preeminent in space science, advanced technology, and large-scale operations. I commented on the Apollo Program and noted that the program would commence with orbital maneuvers and culminate with the one-week trip to the lunar surface. I further stated that for the next five to six years, there would be significant events by which the world would judge the competence of the United States in space.

The summary recognized that in the views of Mr. Webb, Dr. Dryden, and myself, the manned lunar landing, although of highest national priority, would not by itself create the preeminent position we sought. It stated that our future interests in terms of having an adequate scientific and technological base for future space activities demanded that we provide a well-balanced program in all areas, including those not already related to the manned lunar landing.

When the draft was complete, I called Jim at home. He sometimes had severe migraine headaches, and it's not surprising that one occurred at this time. We conferred in his home, with the pages of my text spread across one end of his dining room table. He made few changes until the final summary and conclusions. At that point, he started writing quickly around margins and between lines. For example, he added the following paragraph:

In aeronautical and space research, we now have a program under way that will insure that we are covering the essential areas of the unknown. Perhaps of one thing only can we

37. Transcript of "Presidential Meeting on Supplemental Appropriations for NASA, November 21, 1962," John F. Kennedy Presidential Library, Boston, MA.

38. James E. Webb to President Kennedy, 30 November 1962, Robert Channing Seamans, Jr., papers, MC 247, Institute Archives and Special Collections, MIT Libraries, Cambridge, MA.

be certain, that the ability to go into space and return at will, increases the likelihood of new basic knowledge, on the order of the theory that led to nuclear fission.

I felt that this statement was a bit of a stretch and wondered about Jerry Wiesner's reaction. In addition, Jim put in a plug for the 1964 budget ($6.6 billion), and in the end, he bowed low by saying that if the President felt we should go for the supplemental in 1963, we would give our best effort to its adoption. And it was signed, "With much respect, believe me. Sincerely, James E. Webb, Administrator." The full text of the letter is printed at the end of this book in appendix 2.

1962: Progress, 1963: Objective

I believe that the letter was an all-encompassing summary of policy and programs as viewed by NASA management in the Kennedy years. Remarkably, there was no further discussion of these issues until President Kennedy visited Cape Canaveral a year later. Whether from agreement, exhaustion, or diversion, President Kennedy gave tacit approval to NASA's programs and policies by not engaging with us in further discussions on the questions of NASA's top priority. Preeminence in space on all fronts was our goal; landing men on the Moon within the decade was the top (DX) priority. We were riding two horses.

The year 1962 might be termed the year of the mission or perhaps the year of the spacecraft. After considerable angst between NASA and the White House, Mr. Webb announced the decision that NASA would utilize Lunar Orbit Rendezvous as a means for transporting men to the Moon and back. The development of the Apollo capsule with its supporting service module was well under way at North American Aviation, and Grumman had initiated the design of the Lunar Module for descent to and ascent from the lunar surface. Rendezvous with the Apollo capsule, with its Earth-reentry capability, was obviously essential.

As NASA took stock in 1963, there were three manned Mercury orbital flights to their credit. In addition to John Glenn, Scott Carpenter and Wally Schirra had successfully orbited, and only one flight remained—or would there be two? *Faith 7* (Mercury 9) was scheduled for May 1963 with

Gordon Cooper at the helm. The objective of Mercury 9 was to extend the time in orbit from 6 hours to a day, a result already achieved 21 months earlier by cosmonaut Titov in Vostok 2.

Thursday, 9 May 1963, *Faith 7* (MA-9), Gordon Cooper

Liftoff occurred at 8:00 a.m. Sixty seconds upward, Gordo felt the oscillation of max-q and the rate gyroscopes were giving readings from pin to pin caused by the violent oscillations of the spacecraft. The flight smoothed, and at 3 minutes, the cabin pressure was automatically sealed. Cooper reported, "*Faith 7* is all go." For the next 2 minutes, the Atlas sustainer rocket performed perfectly, and at Mission Control, Schirra reported, "Smack dab in the middle of the go plot." "Beautiful," Cooper replied, "working like advertised." From Guaymas, Mexico, after one orbit, Grissom announced, "Go for seven orbits." As Cooper raced over the launch site, Schirra complained, "You son of a gun, I'm still higher and faster, but I have an idea you're going to go farther." It was an auspicious start.

By the third orbit, Cooper had checked his 11 planned experiments. He prepared to eject a 6-inch sphere with a xenon strobe light. Cooper kicked the switch but couldn't see the flashing light in the dusk or nighttime. However, he did spot the beacon on the fourth orbit at sunset, and on the fifth and sixth orbits, he also saw it flashing.

He ate some bite-sized brownies and fruitcake, kept up with his exercises, took oral temperature and blood-pressure readings, and produced urine samples periodically. The highest priority experiments were the aeromedical.

At the seventh orbit, he was pursuing radiation experiments and transferring urine samples between tanks. The hypodermic syringes were unwieldy and leaked. He placed a message on the tape, "The liquid has to be forced through the piping."

After 10 hours, Cooper was officially informed that he was to go for 17 orbits. The flight had become routine. He spent his last orbit before a rest period having a supper of powdered roast beef mush and some water. He took pictures of India and Tibet, then prepared for a power-down so that he could drift and dream. He was advised by the telemetry command ship *Rose Knot Victor* near Pitcairn Island to "settle down for a long rest." Sometime later, he did relax and fall into a sound sleep; he awoke after an hour, and for the next 6 hours, he napped, took pictures, and taped reports.

On the 16th orbit, Cooper took zodiacal light photographs for University of Minnesota scientists and snapped horizon-definition imprints around the clock for MIT researchers. This latter information was needed for the design of the Apollo guidance and navigation system. On orbits 17 and 18, he took infrared weather photographs of good quality. He resumed the Geiger counter measurements for radiation and continued his aeromedical duties.

On the 21st orbit, a short circuit occurred in a buss bar serving the 250-volt main inverter. The automatic control system was without power. Cooper noticed that the carbon dioxide level was rising in the cabin and in his suit. As he said to the ground controller, "Things are beginning to stack up a little." He completed his checkout and prepared for a manual landing. Glenn gave him the 10-second countdown, and the retrorockets were fired on the mark. Glenn reported, "Right on the old gazoo. It's been a real fine flight. Real beautiful. Have a cool reentry." And that he did.

Cooper landed 4 miles from the *Kearsarge*. As an Air Force officer, he requested permission to be hoisted aboard the Navy's carrier. He next went through arduous medical, technical, and operational debriefings aboard ship and back at the Manned Space Center. He had lost 7 pounds, but after drinking a few gallons of water, he was fine mentally and physically. Cooper proved that man is a pretty good backup for all the equipment on the spacecraft.

My wife Gene was unavailable, so I took my 10-year-old daughter May with me to the Cape for Cooper's triumphant return to Titusville. There was a parade in which May and I rode with Cooper and his wife in an open phaeton. However, at the press conference, all was not sweetness and light. Were we planning a Mercury 10 mission? This possibility had never been presented to me in detail, but I was dead set against it. We had jury-rigged our way with Mercury and had successfully completed our stated goals. Why risk a mission of several days on a Mercury spacecraft when Gemini was designed for missions of two weeks or more? Gemini was behind schedule and needed more focus by the Manned Spacecraft Center, so why divert attention to a questionable objective? Admittedly, Gemini flights were still several years away, but that was all the more reason to emphasize this successor effort.[39]

Returning to Headquarters, I found events of importance happening in rapid succession. Jim Webb asked me if I'd care to participate in a reception for Gordo following the ticker-tape parade in New York. I demurred because I had an appointment that day with George Mueller from Thompson Ramo Woldridge (TRW). I had met him before at TRW, but now I was going to explore the possibility of his assuming a senior position at NASA. It was a good meeting but without specifics. He seemed interested in joining NASA in a key position, and I thought he'd be a potential replacement for Brainerd Holmes, who had already demonstrated poor judgment in pressing the case for supplemental appropriation. However, the jury was still out on his long-term reliability.

Not long thereafter, a luncheon was held in the elegant reception facilities of the State Department. Many individuals were congratulated for their contributions to the success of NASA's Mercury program. It was a love-fest for all but Brainerd Holmes. He called me afterward in high dudgeon because although he ran manned space flight, his name hadn't been mentioned. Well, I thought it would have been better if he had been recognized, but I wasn't in charge of naming honorees. I pointed out that there were individuals like Francis W. Reichelderfer, head of the Weather Bureau, whom NASA wanted to recognize, rather than always patting itself on the back. He said, "Well, I'm certainly not going to Mr. Webb's lawn party." (Mr. Webb had invited many of those attending the luncheon for an informal outdoor reception at his house.) I suggested that Brainerd rethink his priorities, and he did appear, but he had already stepped beyond the bounds. His ego had done him in. He had been out of line in going public on the need for a supplemental, as already discussed. I was concerned about the impact of his present actions on the organization, so I met with Hugh and Jim. We decided that

39. Swenson, Grimwood, and Alexander, *This New Ocean*, p. 495.

we'd had enough, and Jim called Art Malcarney, Brainerd's former boss, who immediately offered to help Brainerd if we felt his role at NASA was irretrievable. We did, and later that day, I advised Brainerd that it was time for him to resign. He didn't question the decision and was glad Art Malcarney would help him back into the industrial world. He resigned on 12 June 1963. In his two years at NASA, Brainerd had placed NASA's manned space program on a fast track, and he returned to a successful career in industry.

Sunday, 16 June 1963, Vostok 6, Valentina Tereshkova

On 14 June, Valeriy Bykovskiy went into orbit aboard Vostok 5 for nearly five days. Then, two days later, on Vostok 6, Valentina Tereshkova became the first woman to orbit Earth. Korolev had a way of informing us that we still had strong competition. Tereshkova is shown in figure 12, along with Yuri Gagarin and Aleksey Leonov. The United States didn't send a woman into orbit until the Shuttle program, over 20 years later.[40]

George Mueller up to Bat

George Low and Joe Shea had been loyal, competent members of Brainerd's Headquarters team. It was essential to keep them involved during the interregnum until Brainerd's replacement was in office, and thereafter as well. I had become deeply concerned about the slow pace of the Saturn I's development. There had been four completely satisfactory launches of the Saturn I first stage. Why hadn't more significant missions been planned to capitalize on that success? Or, as members of the press were unkind enough to ask, why were we spending hundreds of thousands of dollars placing Canaveral sand and water into space? A review of the launch schedule was definitely in order. I worked closely with George and Joe on this and other issues.

On a personal note, my brother-in-law, Caleb Loring, and I owned a series of boats together. Since joining NASA, I had little time for sailing, but we

had planned to race our yawl, *Serene*, from Marblehead to Halifax, Nova Scotia, early in July. The race takes two to three days, and then, after reprovisioning, the return to the Maine coast requires another four to five days. I'd wondered about leaving NASA in its somewhat perilous state, but Jim Webb had convinced me that we all need rest and relaxation (R&R), and the race had been planned with others months before. And so it happened; we left Nova Scotia in thick fog that lasted until we approached Rocque Island, off the Maine coast, on a sunny morning. We were soon spotted by the crew aboard a fleet of yachts who hailed us and said that Mr. Webb awaited my phone call. (Ship-to-ship communications could be picked up by anybody tuned to the particular frequency, and the responses were also on an open circuit.) One of the motor-sailers in the fleet had a high-powered radio telephone, so I climbed aboard and called Washington. Jim had left a phone number to call so that when we discussed George Mueller, we had, in effect, a private line. Jim told me that he and Hugh had interviewed George Mueller and felt he was qualified to manage the manned space program. I readily agreed. George Mueller was sworn in as Holmes's successor on 23 July.

After George arrived, he soon realized that Headquarters was thinly staffed with competent senior managers. He arrived with a list of six or seven officers with whom he had previously worked. Key was Major General Samuel Phillips, who had managed the Minuteman development under General Bernard Schriever. He was extremely competent, and I wasn't certain that the Air Force could or would spare this talent. However, assignment of military personnel to NASA was recognized by Congress as desirable and was permitted for limited numbers in congressional legislation. General Bozo McKee had been Vice Chief of Staff under General LeMay, and he was now on Jim Webb's staff. George's recommendation was reviewed by Bozo and Jim. They readily agreed, and thanks to Bozo, the transfers soon took place. It was a tremendously important benefit to NASA.

It can be said that George was not a person to accept past decisions as a given. In particular, he decide to review the flight schedule, tasking two old

---

40. *Astronautics and Aeronautics, 1963: Chronology of Science, Technology, and Policy* (Washington, DC: NASA SP-4004, 1964), p. 244.

Figure 12. Three cosmonauts: Gagarin, the first in space; Tereshkova, the first woman; and Leonov, the first to perform an extra-vehicular activity.

hands, John Disher and Del Tischler—the first knowledgeable in spacecraft, the second in rockets and launch vehicles. George wanted an unbiased, fresh look. Two weeks later, George was appalled by their findings. They estimated a late 1971 date with a launch within the decade only at unacceptable risk. They immediately came to my office for a similar briefing. When it was over, I took George aside and told him to bury the Disher-Tischler review and create a new program with an acceptable outcome. George was prepared to put forth a radical plan for this purpose. The plan was managerial as well as procedural, and it took into account Congress's reduction of the President's 1964 request from $5.7 billion to $5.1 billion.[41]

We were perilously close to losing control of the program, which placed George in the driver's seat. First, he insisted that the three major Centers in Houston, in Huntsville, and at the Cape should report directly to him. Second, internal to this group, George was the chief executive officer and chairman of the board. The Gemini director, Chuck Mathews; Apollo director, Sam Phillips; and Apollo Application director, E. Z. Grays, all reported to him. At their management reviews, he was the board chairman, with Gilruth, von Braun, and Debus as board members. The project directors were in the Centers. For example, Joe Shea moved to Houston, where he became director of the Apollo capsule and the Lunar Module. There would be no counterparting. Sam Phillips didn't have an Apollo capsule staff person in his office; he worked directly with Joe Shea, as can be seen in figure 30. Each program director had five staff officers, as did the project directors. These were responsible for program control, systems engineering, test, reliability and quality, and flight operations. The two staffs worked closely together. Reductions in the budget included eliminating four Saturn I manned orbital missions. And, of course, the Nova vehicle was no longer required. But these reductions alone would be insufficient to control the budget and to achieve a landing within the decade.

A step-by-step approach adding elements in sequential flights coupled with repeated flights of the final configuration would not, according to George, build reliability into the system. The best

opportunity for reliability and success would come from careful design, redundancy where possible, component quality control, and systems testing. I agreed. As the stages were assembled one by one and then coupled with the spacecraft, the system would be checked out by the same instruments, monitors, and people that would be responsible for the go-ahead on launch day. Finally, on launch day, in the 2-minute period prior to ignition, all key items would be automatically checked to be certain that the readings were within prescribed tolerances. With this procedure, it is only sensible to plan on success. If the first stage is going to do its job, have the second included, and so on, until an all-up systems test is achieved on the first attempt. I was present at Launch Control when Saturn V was launched the first time. Not only did all launch systems perform flawlessly, but Apollo and the service module did also. The Lunar Lander was not yet available, but it would have been included if checked out.

Now, George created quite a stir with his revised program. Words like *impossible*, *reckless*, *incredulous*, *harebrained*, and *nonsense* could be heard behind the scenes. After announcing the plan to the manned spaceflight team, George followed up immediately with detailed schedules. George didn't sell; he dictated—and without his direction, Apollo would not have succeeded.

Kennedy's Last Visit to the Cape

One evening in mid-November 1963, just before leaving my office, I received a call telling me that President Kennedy was thinking of a trip to Cape Canaveral. The White House is always careful never to be too explicit about a presidential trip. Following this, I received a call from Major General Chester V. Clifton, military aide to the President, who gave me more details. He said that the President wanted to get a feel for how we were progressing and that he would have about 2 hours. What did we recommend?

Julian Scheer, NASA's public affairs officer, came down to my office, and several of us sketched a map on the blackboard indicating where the President might land, what he might see up close, and what he might fly over. Naturally, we kept Jim

41. Table 4.4, "Requests, Authorization, Appropriations, Obligations and Disbursements," p. 131, and table 4.13, "Funding NASA's Program FY1962," both in *NASA Historical Data Book*.

Figure 13. Dr. Wernher von Braun explains the Saturn I with its hydrogen upper stage to President John F. Kennedy. NASA Associate Administrator Robert Seamans is to the left of von Braun. President Kennedy gave his approval to proceed with this launch vehicle at his first budget meeting with the Agency on 12 March 1961. (NASA Image Number 64P-0145, also available at http://grin.hq.nasa.gov/ABSTRACTS/ GPN-2000-001843.html)

Webb informed. We felt that the President couldn't cover it all without the use of a helicopter because a couple of bridges connecting the Cape with Merritt Island were not yet completed. I called General Clifton back, and there ensued a series of phone calls and discussions of other opportunities for the President while at the Cape, among them a review of the Navy's Polaris missiles.

The next morning, 16 November 1963, the President flew from Palm Beach to the Cape, where he was greeted by Major General Leighton I. "Lee" Davis and Dr. Debus, the respective heads of military and NASA operations at the Cape, as well as by Jim Webb and me. He was accompanied by Senator George A. Smathers of Florida, a good friend of his. The President said, with a smile, "I'm surprised to see you all here so early on a Saturday morning." Then he stepped into an open car with

Jim and General Davis. They drove by the various complexes rather slowly. We joined them inside the blockhouse at Complex 37, where a Saturn launch vehicle was soon to be tested. There was about a 15-minute briefing there with all kinds of models. The President seemed quite interested in what George Mueller had to say. When the briefing was over, he stood up and went over to the models. He expressed amazement that the models were all to the same scale, because the Mercury launch vehicle was completely dwarfed by the Saturn V. This may have been the first time he fully realized the dimensions of future NASA projects.

We then went out to the pad where the Saturn SA-5 (the fifth Saturn I) was undergoing tests. Figure 13 shows President Kennedy with Dr. von Braun, discussing the Saturn and its dimensions. Before leaving, President Kennedy wanted to walk over and stand right underneath the Saturn. This evidently had come up for discussion with the Secret Service the previous day. They hadn't wanted him to get too close to the rocket. But no matter what anybody thought, President Kennedy was going to go and stand under the Saturn. "Now," he said, "this will be the largest payload that man has ever put into orbit? Is that right?"

"Yes," we said, "that's right."

He said, "That is very, very significant."

I then climbed into the President's helicopter, which had been flown down from Washington on a transport plane for his use. My job was to sit with him as we flew over the new construction area on Merritt Island and to point out the future locations of such things as the Vertical Assembly Building and the launchpads (Complex 39). At the time, 1,800 pilings had been driven into the sand to support the assembly building. Afterwards, we flew about 50 miles offshore to watch a live test of a Polaris missile. Admiral I. J. Gallatin, who was in charge of the Polaris program, described what the President was about to see, which led to a discussion of the whole concept of nuclear submarines, a classified matter about which the President was clearly interested and knowledgeable. We landed on the deck of a waiting ship. The President hopped out vigorously. In honor of his visit, he was presented with a Navy jacket, which, as a naval hero of World War II, he happily put on. He was obviously enjoying himself.

Then, as planned, President Kennedy gave the order to fire the missile. There was a countdown . . . then a hold! I could feel the tension in the Navy personnel there, and I also noticed a couple of Air Force and Army men winking at each other. The President stood watching with binoculars. Fortunately, another Polaris missile was on hand, and the launch was shifted to the backup. When the missile breached the water, we could see that it had "Beat Army" printed on its side. We got back in the helicopter, and the President wore his Navy jacket for the rest of the trip.

On the way back, he brought up the matter of the Saturn SA-5. "Now, I'm not sure I have the facts straight on this," he said. "Will you tell me about it again?" I explained (among other things) that the usable payload was 19,000 pounds, but that we actually would have 38,000 pounds in orbit.

"What is the Soviet capability?" President Kennedy asked. I told him that their payload weighed less than 10,000 pounds. "That's very important," he said. "Now be sure that the press really understands that this vehicle has greater capability than the Soviets'. In particular . . . ," he said, mentioning one reporter by name. Just before we landed, he called General Clifton and said, "Will you be sure that Dr. Seamans has a chance to explain to . . ." (here he mentioned the reporter's name again).

We got off the helicopter and walked quickly over to the President's plane. He shook hands with Jim and the others, then turned back to me and said, "Now you won't forget to do this, will you?" I said I would be sure to talk to the reporter. "In addition," he said, "I wish you'd get on the press plane we have down here and tell the reporters about the payload."

"Yes, sir," I answered. "I'll do that."

Six days later, on Friday, 22 November, I was holding a meeting in my office when I got a call from Nina Scrivener, Jim Webb's secretary. She said, "Something dreadful has happened in Dallas. You'd better come on up to Jim's office."

"You mean the President's been hurt?"

She said, "It may be worse than that."

I closed down the meeting very quickly, then called Gene. "Watch the news," I said. " I don't think we're going to be having that NASA gathering tonight."

Jim Webb had three televisions in his office so that he could have all three networks going at once and flip on the sound of the one he wanted to hear. We sat there watching all three networks. Finally, Walter Cronkite came on and said that the President had died. Gene arrived at the office a little later to distribute the food we had planned to serve at dinner.

We had scheduled our regular monthly program review for the following day. I argued strenuously that we ought to go ahead with the meeting, that President Kennedy had given the Apollo Program a DX priority and that he would have wanted us to press ahead. As a result, we were probably the only federal government organization doing business that day. When the review was over, Jim turned to Hugh and said, "I'm going over to the White House. Do you and Bob want to come along?"

The three of us and our spouses stood in line in the East Room, where the President was lying in state. Everyone wore black. The casket was closed and draped with a flag, with a Marine standing at attention by its side. There was immense grief on every face, and many significant symbols, such as the Great Seal of the United States of America over the door, were draped in black. There were no flowers, no music, only the murmur of hushed voices and the shuffle of feet. It was the saddest place and the saddest time of our lives.

A year after Jacqueline Kennedy Onassis's death in 1994, the *Boston Globe* carried a verbatim transcript of the interview she gave Teddy White soon after President Kennedy's assassination. Teddy White was a well-known historian and author of the book series *The Making of the President*. *Life* magazine requested the interview, and Mrs. Kennedy agreed, provided that Teddy White was the interviewer. He was contacted late in the afternoon in Cambridge and was rushed to the Kennedy compound in Hyannis. *Life* was holding the presses for the article, and so directly after the interview, Teddy dictated the article in Mrs. Kennedy's presence. The article featured her vision of her husband's administration as Camelot and omitted her rather extensive discussion of the space program.

Here are excerpts from the transcript, which appeared first in the Globe on 28 May 1995:

McNamara changed the name at the Cape [Canaveral].[42] Jack was so interested in the Saturn Booster. All I wanted was Jack's name signed on the side of the nose of the booster somehow where no one would even notice. McNamara said that wasn't dignified. But then he changed the name of the Cape itself so that everything that goes to the sky, goes from there.

But I can't see changing the name of something like Sixth Avenue [in New York City]. I don't want to go out on a Kennedy Driveway to a Kennedy Airport to visit a Kennedy School. And besides, I've got everything I want; I have that flame in Arlington National Cemetery and I have the Cape. I don't care what people say. I want that flame, and I wanted his name on just that one booster, the one that would put us ahead of the Russians . . . that's all I wanted.

I'm going to bring up my son. I want him to grow up to be a good boy. I have no better dream for him. I want John-John to be a fine young man. He's so interested in planes; maybe he'll be an astronaut or just plain John Kennedy fixing planes on the ground.[43]

President Kennedy's assassination had a profound impact on the peoples of the world, particularly on those working closely with him in the government. Those responsible for launching the Saturn SA-5, which he had observed and commented on during his inspection in November, wanted some way to express their gratitude for his interest and their grief for his loss. Rumors were rampant that special markings would be placed on the Saturn, which led to the implementation of special security provisions. In the aftermath of the successful launching, while still in the blockhouse, we all felt such an emotional upwelling that there was a near-unanimous request for a call to Mrs. Kennedy. I felt, perhaps wrongly, that such a call would be upsetting for her, and I suggested instead that I carry the sentiments of those involved in the launch back to her in person.

When I returned to Washington, I contacted Walter Sohier, NASA's General Counsel and a friend of the Kennedys'. He didn't think Mrs. Kennedy would be interested in a visit, so imagine his surprise when he and I were invited for tea the following afternoon! Mrs. Kennedy was very gracious, sat patiently as I explained the circumstances of our being there, brought in her children (both recovering from chicken pox), and sent us away exhilarated by our encounter.

My letter to her of 7 February 1964, which follows below, is self-explanatory, but her response of 14 March was unexpected and is deserving of comment. Following her husband's death, Mrs. Kennedy had had little time in which to move out of the White House into a house on N Street,

---

42. Geographic names cannot readily be changed, but Mrs. Kennedy received her wish. NASA's launch facility at the Cape is now called the John F. Kennedy Space Center.

43. Transcript of Teddy White's interview with Jacqueline Kennedy, *Boston Globe* (28 May 1995).

in Georgetown, loaned to her by a friend. Remarkably, she started immediately to reply to the huge number of people who had attended the funeral service or offered their condolences in other ways. My wife Gene was among the many who volunteered their assistance. Hence, Mrs. Kennedy's personal response to my visit and letter is truly remarkable. First, my letter:

February 7, 1964

Mrs. John F. Kennedy
Washington, D.C.

Dear Mrs. Kennedy:

Thank you for the pleasant visit you afforded us Monday evening. It meant a great deal to me to be able to tell you about the recent Saturn launch from the John F. Kennedy Space Center.

The accompanying detailed engineering model of the actual Saturn launched on January 29th is presented to you with appreciation from all of us. It was utilized by Dr. Wernher von Braun and the staff of the George C. Marshall Space Flight Center which has responsibility for the Saturn development.

Having seen young John's interest in space toys, and having barely escaped from your home with the other model that I brought, I am also sending some fairly sturdy launch vehicle models for his enjoyment.

Sincerely,

Robert C. Seamans, Jr.
Assoc. Administrator[44]

Following is her reply:

Mrs. John F. Kennedy

March 14, 1964

Dear Dr. Seamans,

I do thank you for that most precious model of the Saturn—the one that Wernher von Braun and everyone worked on—(I could not believe my eyes when Walter Sohier brought it).

John had a fleeting happy look at it—and then I sent it to Archives—to go in Jack's library.

Your thoughtfulness has touched me so much—that you would wish to come—and tell me about the Saturn booster—and think of calling me from the blockhouse when it was going off. All I care about is that people still remember what Jack did—and you were always thinking of him.

Then when you came and saw John—it was so kind of you to see how a little boy who had grown up so close to a father who always had exciting new plane and rocket models in his office to show him—who took him on his most cherished plane and helicopter rides—would still care so much about all those things—and feel so cut off now that they are no longer a part of his life.

Those "heavy duty" models that you sent him are his joy—taken apart and put together constantly—I do thank you more than I can say, for your thoughtfulness to him and to me—

Sincerely,

Jacqueline Kennedy[45]

---

44. Robert C. Seamans to Jacqueline Kennedy, 7 February 1964, Robert Channing Seamans, Jr., papers, MC 247, Institute Archives and Special Collections, MIT Libraries, Cambridge, MA.

45. Jacqueline Kennedy to Robert C. Seamans, 14 March 1964, Robert Channing Seamans, Jr., papers, MC 247, Institute Archives and Special Collections, MIT Libraries, Cambridge, MA.

Chapter 4:
JOHNSON'S SOLID SUPPORT

Communications for Mission Control

Starting with Gemini 4, NASA controlled manned orbital missions from Mission Control in the Manned Spacecraft Center in Houston, Texas. Up until then, the network had consisted of 12 ground stations and several ships. At each station, there was a full complement of communicators, doctors, and engineers to work with the capsule crew overhead. This duplication was necessitated by the fact that there was no wideband communication back and forth through the network and, hence, to the spacecraft. President Kennedy's first supplemental in 1961 contained $10 million for communication satellites, and $50 million was added to the second supplemental. In the fall of 1961, a sole-source contract was negotiated with Hughes Aircraft for a small, spinning, geosynchronous satellite. Keith had acquainted me

with its proposal on my second day at NASA when he took me to lunch at the White House mess. Now that the funds were available, Hughes's concept, called Syncom, would be put to the test. Incidentally, the $10 million was used for a lower altitude satellite called Relay under contract to RCA. We also had an agreement with AT&T to launch its satellite, Telstar, on a reimbursable basis. However, both Relay and Telstar would require 20 to 25 satellites to obtain world coverage. The second launch of Hughes's small, spinning, synchronous satellite was successful and was adopted by Intellsat, an international communication agency, and Comsat, Intellsat's agent for its first prototype. Comsat adopted the Syncom concept and contracted with Hughes Aircraft for Early Bird, the heart of the first satellite communication system to be commercialized. And Comsat's first major contract for communication services was with NASA. With three geosynchronous Early Birds spaced around the equator, NASA was provided with continuous worldwide broadband communications throughout the Gemini and Apollo Programs.

International Business Machines' 360-75 Computer

Bob Gilruth, the head of the Manned Space Center, came to Washington to inform Headquarters that Mission Control had insufficient computer capacity for lunar operations. The only solution, he maintained, was to buy five International Business Machines (IBM) 360-75 computers for $60 million. At least three other companies had mainframe computers—Sperry, Control Data, and RCA—and if we went sole-source to IBM, those companies would probably complain to the Government Accounting Office (GAO). This action could lead to a GAO investigation, with a delay in entering negotiations with IBM or possibly an upset with a requirement for open bidding. Bob explained that IBM claimed that it had already spent $10 million of its own funds on programming efforts for NASA. We asked Bob if the computers were catalog items. He thought so and believed that the company had already sold a few, but that was insufficient rationale for a sole-source contract.

Jim Webb stepped into the breach. In the course of one week, the CEOs of all four contractors were invited to NASA Headquarters on separate days. They could bring whomever they wished. The situation was explained to each. NASA wanted a fixed-price contract for any computer that would satisfy its operational needs. The delivery date would have to be specified, with penalties for late delivery. Each contractor was free to spend whatever time they needed at the Manned Spacecraft Center to learn NASA's requirements. Both RCA and Sperry bowed out, saying they'd like to be considered for the peripherals. Control Data spent a month at Houston before advising NASA that it was in no position to bid. So IBM received $60 million for five 360-75s, and the contract was upheld.

Gemini Shake-Up

Originally, Jim Chamberlain had directed Gemini, back in the days when it was still Mercury II. During his tenure, the configuration evolved into a two-man spacecraft with a flight capability of up to two weeks. Jim commenced the negotiations with McDonnell, but he became obsessed with the use of Gemini for the manned lunar landing mission. His actions created sufficient diversion at the Manned Spacecraft Center that he was replaced, and Chuck Mathews took charge of the project.

George Mueller inherited the Gemini Program that was to be run by NASA and McDonnell Aircraft in the same fashion as Mercury. On average, Mercury flights took place every five months. With 12 flights planned (two unmanned), it would take five years to complete the program, or stated differently, commencing in 1964, Gemini would not be complete until 1969—hardly in time to be of assistance to Apollo if flights occurred at the same pace as those of Mercury. Examination of Mercury procedures showed George that each capsule was built in St. Louis by McDonnell and tested there by NASA. The capsules were then shipped to Cape Canaveral, where another team of McDonnell and NASA employees tore them down for inspection and returned them after reassembly. The launch team didn't want to be responsible for an unfamiliar or nonworking spacecraft. George decreed that one team would assemble and inspect the capsule in St. Louis, move with the capsule to the Cape, reinspect for shipping damage, mate the capsule with the Titan launch vehicle, and assist in the launch. There was nearly a year's gap between Gemini 1 and 2, but after that, Gemini launches occurred about every two months. Gemini 12, the final flight, took place on 11 November 1966, well in time to provide lessons learned for the Apollo Program.

George Mueller brought a new style to NASA, but he also inherited personnel who were willing and able to follow his lead. This capability was certainly manifested in Charles "Chuck" Mathews, a former member of the research staff with whom I'd worked in days of yore. The purpose of Gemini was to gain operational experience by using Mercury-like assets. The team at McDonnell Aircraft moved smartly into the new scaled-up Gemini configuration, which had a shape and heatshield similar to those of Mercury. However, Titan II could place nearly 9,000 pounds into orbit compared to the Atlas. Mercury could weigh only 4,400 pounds. The expansion in size and weight allowed Gemini to have considerably greater capability than its predecessor Mercury.

Tuesday, 23 March 1965, Gemini 3, Virgil Grissom and John Young

On the initial flight launch, Mission Control was still in the Mercury building at Cape Canaveral, and the world tracking stations were also the same as before. With President Johnson in the White House, Vice President Hubert H. Humphrey became the Chairman of the Space Council. Just before leaving for the Gemini 3 launch, I found out that the Vice President was going to attend with one of his sons. Mr. Webb was strongly opposed to his presence. He felt that there were risks in every space mission and that the Vice President could be placed in an awkward position of major media exposure with insufficient knowledge of the subject.

However, Humphrey was an easy VIP to escort. He said he didn't want special privileges for his son, so I arranged for him to be with my sons, Toby and Joe, at the media briefing. The Vice President and I were in Mission Control. We were seated on a balcony overlooking the professionals at their consoles. There were world charts on the wall showing different orbits and Gemini's progress. Chris Kraft was the Mission Director. Humphrey had trained as a pharmacist in his earlier days, so he was particularly interested in the medical testing of the astronauts and the instruments they carried on their bodies.

After liftoff, I walked the Vice President onto the floor of Mission Control to meet Chris Kraft and his team. In introducing Chris, I said, "Can you imagine naming someone Christopher Columbus Kraft?"

The Vice President responded, "Can you imagine parents naming a son Hubert Horatio Humphrey?" A good time was had by all on the floor of Mission Control. Then there was a hiatus of a few hours before a scheduled press conference following the successful reentry and recovery of the two astronauts, Gus Grissom and John Young.

I escorted the Vice President to the briefing, which, after a successful three-orbit flight of 4 hours and 53 minutes, would be quite pro forma. However, the old warrior was somewhat overwhelmed at the media's size and interest. As we approached the main tent, large, portable generators were throbbing, and as we entered, we were illuminated by many flashbulbs. I started the press conference by welcoming everyone to the successful initiation of the Gemini Program. I introduced the Vice President, who spoke briefly, and then it was on to the flight particulars. Questions were easily handled, and soon it was goodbye to the Vice President and his son.

I stayed at the Cape for the debriefing when the astronauts returned to the Cape the next morning. That morning, by sheer chance, Ranger 9 was plowing into the Moon, and the lunar photographs were displayed live on morning TV. Ranger had been such a trying project for NASA's Jet Propulsion Laboratory (JPL). There had been a series of launch vehicle and spacecraft failures. For example, it was a glorious night in Washington, with a full Moon, when Ranger 6 approached the lunar surface with high expectations. Unfortunately, a relay failed to function, the cameras weren't initiated, and no photographs of the Moon were taken.

The resulting publicity was most unfavorable to JPL. A thorough investigation was conducted, and a variety of equipment and procedural changes were recommended and executed. The final three Ranger missions were entirely successful, so I was most happy for JPL when I viewed Ranger's final detailed lunar photography on a Holiday Inn TV set.

Soon thereafter, I received a phone call advising me of the awards ceremony to take place at the White House in a few days. I was to receive the NASA Distinguished Medal along with Grissom and Young. Whom did I wish to attend? I obviously suggested my wife Gene, our five children, and my mother and father, as well as Gene's sister Romey and her husband Caleb. I heard later that I had overreached a bit, but they all were invited, and all came except for our daughter Kathy and her husband Lou. They had been at Cape Canaveral to see the Gemini launching and had stopped by my motel at 6:00 a.m., but by then I was long gone. They had left no message as to their whereabouts.

I was, of course, both surprised and pleased to be one of the three honorees at the White House ceremony. After the formal awards, the media asked for photographs of the astronauts, their families, and President and Mrs. Johnson. As the Youngs were returning to their seats, Mr. Webb stood up and said, "Don't forget the Seamanses." So we trooped up and had our time in the Sun with the Johnsons. Toby, Joe, and May were quite grown up, but Dan was only six and looked quite startled by the klieg lights. The photos of us all in our local papers had the small boy in the foreground, erroneously identified in the caption as the son of John Young. As Mother said when she sent the clipping, "Doesn't he look a lot like Daniel?" And, of course, he was.

The media loved the astronauts. They were fun, skilled, adventuresome heroes, but Jim Webb thought that others also were making major contributions and should share in the glory. For that reason, Gene and I flew to New York with the astronauts, their wives, and Vice President Humphrey. Unfortunately, ticker tape gets pretty soggy on a rainy day, but we drove up Broadway in open cars for an awards ceremony at City Hall. This event was followed by a meeting with U Thant at the United Nations, a lunch with Mayor Wagner, and a gala reception at the Waldorf. I believe I shook hands with several thousand people, and my tennis elbow was none the worse. With perhaps a little arm-twisting, Jim Webb got his wish to have other professionals honored alongside the astronauts, and I was the beneficiary.

Five days before the Gemini 3 mission, Aleksey Leonov had stepped outside of Voskhod 2 for the first extravehicular activity (EVA). One of the principal reasons for Gemini was to gain experience with EVA. If the Lunar Lander successfully conducted a rendezvous with the lunar orbit capsule but couldn't dock for mechanical or other reasons, the crew who had landed on the Moon could transfer to the return capsule by EVA. The Gemini team advised me that they were prepared to conduct an extravehicular experiment and asked whether I could visit Houston to review their effort.[1]

Thursday, 3 June 1965, Gemini 4, James McDivitt and Edward White II

During my visit, I had an opportunity to experiment with the handheld maneuver unit, which had four small jets several feet apart and aimed in the same direction. There was no combustion; rather, gas forced from the jets provided the thrust. Aiming the jets in one direction, the astronauts would move in the opposite. The testing at Houston was done on an air-supported, friction-free platform, obviously in only two dimensions. Even a neophyte like me could maneuver about quite naturally and easily. I returned to Washington prepared to recommend the EVA test on Gemini 4. Hugh Dryden was quite opposed. He felt that the experiment was jury-rigged in fast response to the Soviets. I wrote a careful memo to Jim stating that we entailed risk every time we sent astronauts into orbit and that there should be an attempt to accomplish as much as possible on each mission. I added that I felt that the team in Houston, including the astronauts, had the necessary equipment and were well prepared for its use. The memo came back to me as "approved."

Gemini 4 was a great success, achieving all objectives and providing photos of Ed White outside the capsule. These photos can still be seen today in public displays. Ed was able to maneuver with the handheld unit to the extent of his tether (see figure 14). After 10 minutes, when asked to reenter the capsule, Ed was lollygagging a bit while expounding upon the sensational view. He then found it more difficult to move himself and his equipment into the capsule than expected, but finally the capsule was secure and repressurized. The astronauts failed to conduct stationkeeping and rendezvous activities with the second stage of the launch vehicle, but they executed prescribed in-plane and out-of-plane maneuvers, as well as 11 small experiments. Mr. Webb and I picked up Ed and Jim at the airport prior to the briefing and awards ceremony at the Manned Spacecraft Center. He invited them to sit with him in the backseat so that he could let them know that their attitude had been too carefree and undisciplined during the EVA. He was obviously displeased. However, all was forgiven at the Center briefing and, later, at the Rose Garden ceremony with the President. The mission also successfully demonstrated the new mission control capability at the Manned Spacecraft Center.[2]

The Oval Office and the LBJ Ranch

One day in late August, I had just played tennis and was in the shower when Gene pulled the curtain aside and said that the President was on the phone. I picked up the phone while dripping water

---

1. Ivan D. Ertel, "Gemini 3 Flight," in *Gemini Program* (Houston, TX: NASA, 1967).

2. Ivan D. Ertel, "Gemini 4 Flight," in *Gemini Program* (Houston, TX: NASA, 1967).

Figure 14. On 3 June 1965, Edward H. White II became the first American to step outside his spacecraft and let go, effectively setting himself adrift in the zero gravity of space. For 23 minutes, White floated and maneuvered himself around the Gemini spacecraft while logging 6,500 miles during his orbital stroll. (NASA Image Number 565-30431, also available at http://grin.hq.nasa.gov/ABSTRACTS/GPN-2000-001181.html)

on the carpet. The White House operator said, "Just a minute, the President."

On he came: "Good morning, Bob, I just wondered if you would care to visit me today on my birthday?"

I answered, "Yes, sir." I arrived at the White House for our noon meeting but didn't enter the Oval Office until an hour later. When I entered, the President was reading the news on his ticker tape and finally turned to me.

"Seamans, sit over there." He then proceeded to say, "Seamans, you guys at NASA are good with your science and know how to work the Hill, but for me you're a great big zero." As he made a large

zero with his forefinger, he asked, "What do you think is most important to me?" And then he gave the emphatic answer, "It's peace, Seamans, it's peace!" I began to see the light. Politicians love to include astronauts in community and even worldwide events. We had Gemini 5 in orbit with Pete Conrad and Gordo Cooper aboard. But the President didn't tip his hand. He said, "I want you to call within the hour and tell me that Jim Webb will at my ranch in Texas to go to church this Sunday morning with me and Dean [Rusk]." I told him that Jim was having a well-deserved weekend in North Carolina with his family. His response was clear: "Jim is the best damn administrator in Washington; tell me within the hour he'll be there." Fortunately, I reached Jim, who said he'd be there, but he wanted me to be there as well. When I called the President and told him Jim's wish, he said, "Why, of course, Bob, come too and bring your family." I thanked him and said that only Jim and I would be present.

Saturday, 21 August 1965, Gemini 5, Gordon Cooper and Charles "Pete" Conrad, Jr.

We arrived at Johnson's ranch Saturday evening and had a meeting with the President and his press secretary, Bill Moyer. The President discussed his plans for the astronauts to visit Greece, Turkey, and several African countries. He then had Bill Moyer read his press release for Sunday morning and asked for comments. After some discussion, we concurred, and Sunday morning before church, the President sat outdoors with the media present and called the astronauts to congratulate them on their eight-day trip. He also advised them of their upcoming goodwill trip.

Gemini was in orbit for nearly eight days, the longest duration manned flight to date. During the flight, the astronauts completed a wide variety of medical and observational experiments. Six were for the Department of Defense. A Radar Evaluation Pad was ejected for radar checks, and for the first time, a fuel cell was used as the source of electrical power.[3]

Mission 76: Saturday, 4 December 1965, Gemini 7, and Wednesday, 15 December 1965, Gemini 6

The 76 mission involved two spacecraft. Gemini 7 was scheduled for 14 days in orbit, with Frank Borman and Jim Lovell as crew. Originally, Tom Stafford and Wally Schirra were to rendezvous and dock with an Agena unmanned capsule. When the Agena failed to get into orbit, the mission was scrubbed and replaced with a mission for rendezvous with Gemini 7 instead. Since Titan II had only one launchpad, only one week instead of the usual two months was available to clear the pad after Gemini 7's liftoff and refit the facility for Gemini 6. Time had to be allowed for erecting the launch vehicle, mating it with the capsule, and checking out the assembly. The schedule for the launchpad was programmed hour by hour, night and day, for the full week. I reviewed the plans, and they appeared tight but feasible. Frank Borman and Jim Lovell lifted off on 4 December and had been in orbit eight days when, on Sunday morning, 12 December, engine ignition commenced and, within a second, was turned off. No one understood the reason why. It turned out that a plug that was supposed to open at liftoff fell off a few seconds early due to the engine vibration. President Johnson called Jim Webb an hour later and said that he was very disturbed by the failure. When Mr. Webb called me, I said that the President should be very proud of the cool heads that didn't panic. At ignition, all pyrotechnics are armed for separations, the umbilical arms are separated from the vehicle, and the fuel and oxidizer are flowing. An explosion might occur. The crew could have ejected. Instead, they worked with the ground crew and followed planned procedures, and there was still a chance for a successful mission.

Gemini 6, with Wally Schirra in command and Tom Stafford as pilot, was successfully launched on 15 December. They conducted their orbital maneuvers step by step. After 5 hours and 15 minutes, Gemini 6 was slightly below and 37 miles behind Gemini 7. Tom Stafford flew Gemini 6 to within 2 feet of Gemini 7 and then proceeded to inspect Gemini 7 from all directions (see figure 15). Gemini 7 then did the maneuvering. All hands took turns controlling the vehicles, and then the spacecraft were parked 30 miles apart so that all could sleep. Much valuable data information came from what became known as Mission 76. Foremost were the medical information from the 14 days in flight and the orbital experience of rendezvous and formation flying. Many awards were made and were rightly deserved by the astronauts, Deke Slayton, and the facility technicians. (Deke was one of the first seven astronauts; he was grounded by a heart defect and became the astronauts' manager.) However, actual docking was still to be achieved.[4]

The Loss of Hugh Dryden

On 2 December 1965, Hugh Dryden died after a protracted battle with cancer. President Johnson said, "The death of Dr. Dryden is a deep personal loss. No soldier ever performed his duty with more bravery and no statesman ever charted new courses

---

3. Ivan D. Ertel, "Gemini 5 Flight," in *Gemini Program* (Houston, TX: NASA, 1967).

4. Ivan D. Ertel, "Gemini VII/Gemini VI–Long Duration Rendezvous Missions," in *Gemini Program* (Houston, TX: NASA, 1967).

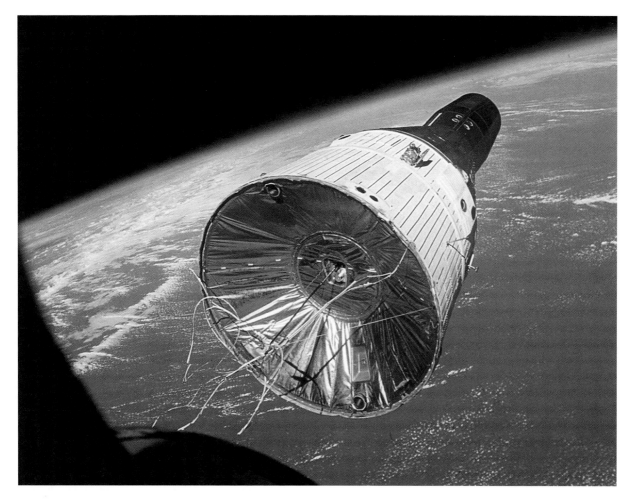

Figure 15. This photo of the Gemini 7 spacecraft was taken through the hatch window of the Gemini 6 spacecraft during rendezvous. (NASA Image Number S65-63221)

with more dedication." Vice President Humphrey was most prescient when he noted, "We shall miss him [Dr. Dryden] sorely as we plot our course for the decade ahead."[5] I had known Hugh since 1948, when he was the director of the NACA and I was appointed to the Subcommittee on Stability and Control. Our subcommittee had sufficient controversy that we met with Hugh on occasion. He was a master at resolving contentious issues.

At NASA, I briefed him many times and gratefully received his wise counsel, both technical and political. When the Instrumentation Laboratory obtained the contract for Apollo's guidance and control, I received a letter from Dr. Charles Stark "Doc" Draper advising me that he always observed first-hand the performance of his developments, whether

in submarines, surface ships, bombers, or fighters. So he wanted me to know that he was ready and eager for a spaceflight. Doc was for many years the head of MIT's Department of Aeronautics and Astronautics and the director of its Instrumentation Laboratory. I knew that as much as we would appreciate his hands-on advice, the rules for flight were stringent, and he didn't qualify. At the next meeting with Webb and Dryden, I read Doc's letter. Jim Webb was ecstatic: "I'm going to show the President the letter this afternoon. The letter shows how some scientists truly believe in manned spaceflight."

However, Hugh piped up, "Before you go, have you considered all the implications? One in 20 applications for astronaut duty is accepted. Many fail for medical reasons. Doc is 30 years older than most

5. *Astronautics and Aeronautics, 1965: Chronology of Science, Technology, and Policy* (Washington, DC: NASA SP-4006, 1966), pp. 534–535.

astronauts." Jim reluctantly reined in his enthusiasm. Hugh had saved NASA from an embarrassing incident. Hugh was a wise, moderating influence on many occasions.

Soon after Hugh's death, I received an interim appointment as Deputy Administrator, and I was sworn in on 21 December. Then, on 2 January 1966, changes in the organizational structure were announced. The Office of the Administrator would now include the Administrator, Jim Webb; the Deputy, me; the Associate, also me; the Associate Deputy, Willis Shapley; and the Executive Secretary, Colonel Lawrence Vogel. Larry Vogel became responsible for channeling and scheduling work within the office. Willis Shapley became the principal assistant to both Jim Webb and me. I became Mr. Webb's alter ego at the same time that I continued to be the general manager of all NASA operations. Fortunately, George Mueller, Homer Newell, Mac Adams, and Edmund Buckley continued as managers of their respective programs. I was confirmed by the Senate on 29 January 1966 and was sworn in again by Jim Webb in the presence of the Vice President. Also on that date, Mr. Webb and Mr. McNamara signed a Memorandum of Understanding for a joint Manned Space Flight Policy Committee to be chaired by John Foster, Director of Research and Exploration for the Department of Defense, and me. These arrangements superseded the NASA-DOD Gemini understanding of 21 January 1963. My plate was getting fuller, and I would have less time in the future to follow individual projects actively. I truly missed Hugh Dryden.

Wednesday, 16 March 1966, Gemini 8, Neil Armstrong and David Scott

Gemini 8 was most significant, most fraught with danger, and most skillfully managed. Rendezvous and docking were, of course, key to the success of the Apollo lunar landing mission. Without a successful rendezvous, the two astronauts who landed on the Moon would not be able to return to Earth. During the preparation phase for Gemini 8, I spent time in Houston and had an opportunity to ride with Neil Armstrong in the rendezvous and docking

simulator. The Gemini simulator in which we were inserted could yaw, pitch, and roll as well as translate three dimensionally. Neil negotiated the final stages of rendezvous and then successfully docked with the target. When asked to try my skills, I demurred—better not to risk damage to a key facility. I was impressed and waited for 16 March with eager anticipation.

The Agena lifted off from Cape Canaveral on schedule at 10 a.m. eastern standard time (EST), followed as planned by Neil Armstrong and Dave Scott aboard Gemini 8 at 11:41. They performed a coplanar change of half a degree and a series of apogee and perigee adjustments, finally achieving a radar lock on the Agena when the Gemini was 158 miles astern. They achieved a visual sighting at 76 miles, 4 hours and 40 minutes into the flight. The Gemini was brought within 60 to 80 feet of the Agena for a visual inspection. All appeared in good order. Neil Armstrong then prepared for final alignment and docking. He closed to within 2 feet of the docking adapter. He was over the eastern Pacific, in communication with the tracking and communication ship *Rose Knot Victor* below. Docking was approved, and the Gemini proceeded toward the Agena at 0.75 mph until the Agena-Gemini was latched together. The time was 6:15 p.m.[6]

Soon after this historic moment, I left home elated to attend the Goddard dinner where 1,500 of the aerospace fraternity would be dining and celebrating this most recent success. Imagine my surprise when, upon arriving at the hotel, I was whisked away to a private room. I was told that the Gemini had had to separate from the Agena and was spinning at an uncontrolled rate. I mistakenly decided to attend the dinner as planned. I should have returned to my office, but I did arrange for continued surreptitious updates. In the course of the dinner, I first told the crowd that the Gemini spacecraft was in a perilous state. Some thought I was joking, but then there was stunned silence. Later, I could announce that Gemini was de-spun and that an emergency recovery would be made in the Pacific, 500 miles east of Okinawa. The destroyer USS *Mason* and two C-54 aircraft were converging on the area.

---

6. Ivan D. Ertel, "Gemini VIII—Rendezvous and Docking Missions," in *Gemini Program* (Houston, TX: NASA, 1967).

It was agreed that the Vice President would deliver his speech and I would give him the high sign when the astronauts were safe. The Vice President was noted for his gift of speech, but even he was fading when I finally gave him the okay sign. The Gemini had been spotted by an aircraft. Of course, pickup remained, and Jim Webb felt that I had taken undue risk in making the announcement. But the guests left reassured after an agonizing dinner.

The Gemini landed at 10:22 EST, and Pararescue men dropped from a plane 13 minutes later. The flight crew was picked up by the *Mason* early the next morning at 1:28 EST, and 9 minutes later, the spacecraft was aboard. The destroyer docked in Okinawa 18 hours later. Ultimately, five aircraft were involved in the rescue.

Although the mission was nearly a disaster, there were many significant achievements:

- The first successful simultaneous countdown and launch of two vehicles on the same day at the precise minutes planned

- The successful retrieval of spacecraft and astronauts in a secondary landing area required for the first time

- The first docking and maneuvering of two vehicles in space

Tape analysis showed that the errant yaw thruster had spun the capsule after separation from the Agena at nearly one revolution per second. Neil and Dave successfully diagnosed their predicament, turned off all thrusters, and utilized the reentry control system to de-spin. During the emergency period, as the crew members were undocking, they had the presence of mind to leave the Agena responsive to ground command and with the tape data intact so that readout on the ground was possible—extremely important for the accident review team. In addition to the astronauts, many others deserved great credit. Special commendation was given at the postflight news conference to General Huston on behalf of the DOD recovery team; to Admiral Persons, commander of Task Force 130, to which the USS *Mason* was assigned; and to the three Pararescue men who attached the floating collar around the Gemini.

Friday, 3 June 1966, Gemini 9, Thomas Stafford and Eugene Cernan

Tom Stafford was in command of the Gemini 9 mission, and Gene Cernan was the pilot. Theirs was the hard-luck mission, but through no fault of their own. As in Gemini 8, the Agena was to depart an hour and a half ahead of the astronauts. Unfortunately, only 2 minutes into its flight, all contact with the target vehicle ceased. Shortly thereafter, George Mueller announced that the Augmented Target Docking Adapter (ATDA) would be used by Gemini 9 for rendezvous trials.

The ATDA liftoff on Atlas proceeded on schedule, but the guidance system update couldn't be transferred through the ground equipment to the spacecraft. When two other attempts failed, the flight was scrubbed and rescheduled for 3 June, two days hence. On the third, all systems were go, liftoff was nominal, and the spacecraft was soon in orbit. After a series of orbital corrections, the astronauts were in sight of the adapter. To their dismay, the shroud on the nose of the ATDA had not separated. Stafford reported that both the clamshells of the nose cone were still on but were open wide. It looked like an angry alligator. After close-proximity inspection and two additional rendezvous approaches, the crew was tired and put off EVA, the second objective of the mission, until the morrow.

Gene Kranz, the Mission Control Flight Director who later became famous for controlling the return of the crippled Apollo 13 and his watch cry of "failure is not an option," was advised by the astronauts that they were ready for depressurized action. The word was "go" from the tracking station at Carnarvon. This activity took place between Canton Island and Hawaii, culminating in an open hatch with Cernan standing and prepared to egress. He gradually worked his way to the adapter section with the "snake" (umbilical) all over him. He then reloaded the EVA camera and turned on the EVA lights for nighttime operations. Back at the adapter area, he began to plug into the Astronaut Maneuver Unit. He started noting fogging in his helmet visor. As he proceeded, Stafford commented that Cernan was finding his work four to five times more difficult than in ground test, so Stafford and Cernan evaluated the situation after sunrise and felt that the fogging constituted a flight safety hazard. The experimental maneuver unit was scrubbed. By resting, Cernan gained back 25 percent of his vision, but as he began to retrieve equipment, the fogging grew worse again. The flight plan called for photographs at sunset, but Stafford decided to bring Cernan back into the spacecraft. Cabin repressurization occurred without incident. Gemini 9 landed less than 2 miles from the USS *Wasp*. At the news briefing two weeks after the flight, it was noted that the liftoff and reentry were flawless. It was stressed that the Gemini Program was experimental and that on each flight, an attempt was made to advance our understanding of space operations to the maximum extent possible. Clearly, much experience was still needed before Apollo.

The Gemini Program concluded with three successful missions. The three were similar in objectives and operations, and each helped clear uncertainties, especially difficulties encountered in astronaut activities external to the capsule. The question of why EVA was so much more strenuous in space then in preflight training would ultimately be answered and the trouble corrected.[7]

Monday, 18 July 1966, Gemini 10, John Young and Michael Collins

The Agena was launched from Complex 14 by the Atlas booster at 3:40 p.m. EST, and John Young and Mike Collins were to follow at 5:20 p.m. within a 37-second window. Gemini 10 lifted off exactly as planned. Through a series of orbital maneuvers, the astronauts brought their spacecraft within 40 feet of the Agena and were given a "go" for docking from the *Coastal Sentry* tracking ship. John executed the docking maneuver with precision, and the Agena pulled the nose in to make a rigid connection between the two vehicles.

Using the primary propulsion system of the Agena, Collins conducted a series of three maneuvers. In the first, the Gemini was driven forward at one g for 1 minute and 24 seconds. This thrust increased the apogee of the orbit to 412 nautical miles, a record for manned flight and one that provided a magnificent view of the Middle East. After

two more Agena burns, the orbit was a little over 200 nautical miles in altitude.

The EVA lasted for 50 minutes. After depressurization of the capsule, Mike first stood up in the cockpit and gazed with wonder as the world sped underneath him. He then used the maneuver unit to visit the Agena and collect a data package fastened there.

The astronauts returned to Earth after three days, within 3.4 miles of the intended landing point. They were retrieved by the recovery ship USS *Guadalcanal*. At the awards ceremony, I noted that Gemini was proving to be much more than a two-man Mercury. I said that Gemini had given us the ability to change orbits, inspect other spacecraft, rendezvous and dock, and use Agena as a switch engine in space.[8]

Monday, 12 September 1966, Gemini 11, Charles "Pete" Conrad, Jr., and Richard Gordon, Jr.

Gemini 11, with Pete Conrad and Dick Gordon, performed a most difficult maneuver, namely, to rendezvous and dock with the Agena on their first orbit. The Atlas-Agena had lifted off at 8:05:02 a.m. EST, and Gemini 11's liftoff had to be at 9:42:26. Their window was only 2 seconds. Although successful, Pete said he wasted 2 or 3 percent of his fuel fussing around and getting used to seeing the bright Agena when he couldn't see his instruments. He said he fumbled around trying to find his glasses. Later, with his sunglasses, he had no trouble reading the instruments and tracking the Agena.

The following day, the Gemini-Agena configuration reached an apogee of 742 nautical miles and, in the ensuing orbits, performed a series of photogenic experiments of terrain, weather, and airglow horizon. Afterward, they used the Agena to return to a nearly circular orbit at 160 nautical miles of altitude.

The astronauts achieved the greatest total time in extravehicular activity, a total of over 2.5 hours. However, Gordon experienced overheating and fatigue, as had others in previous missions. After opening the hatch, he set up a camera, received an experimental data package, and attached a tether to the Agena docking bar. This phase of EVA was ter-

7. Ivan D. Ertel, "Gemini IX-Rendezvous Missions," in *Gemini Program* (Houston, TX: NASA, 1967).

8. Ibid.

minated because of pilot fatigue and the fact that perspiration was gathering in the pilot's right eye and impairing his vision.

However, the astronauts were able to conduct a tethered exercise with Gemini and Agena roped together and rotating at 55 degrees per minute. Although very small, this was the first artificial gravity achieved in space.

Retrofire occurred over Canton Island, and the landing point was achieved automatically for the first time. Splashdown happened within 1.5 miles of the USS *Guam*. At the press conference, General Davis, commander of the Atlantic Missile Range, praised the precise landing and noted that recovery was becoming an easy assignment. George Mueller had opened the press conference by listing the many major accomplishments of Gemini 11. Bob Gilruth found the results with the tether to be fascinating but closed by emphasizing the hard work that had gone into preparing for extravehicular operations—and the fact that, as yet, this effort had not yielded satisfactory solutions.[9]

Friday, 11 November 1966, Gemini 12, Jim Lovell and Buzz Aldrin

On 11 November 1966, with Jim Lovell and Buzz Aldrin aboard, Gemini 12 lifted off from the Cape within half a second of its prescribed time in hot pursuit of Agena. The astronauts followed a series of maneuvers in order to rendezvous on the third orbit. They achieved an early lockon at 235 miles, but as they approached the Agena, the radar failed, and the rendezvous took place using backup visual procedures. Docking and a variety of orbital maneuvers took place with and without Agena, but the primary emphasis was on Buzz Aldrin and his extravehicular performance. The first day, he spent 2 hours and 29 minutes in standup activities; on the second day, 2 hours and 8 minutes; and on the third day, 51 minutes while standing in the open cockpit. During the EVA, Buzz utilized portable handrails, foot restraints, and waist tethers. This latter system consisted of two tethers attached to the astronaut's parachute harness. The tethers were hooked to rings, for example, on the Agena docking cone and

on a telescoping handrail. Aldrin said, "With this system, I could ignore the motion of my body and devote my full effort to the task at hand." He also found that the underwater simulation of EVA had provided his most helpful training.

On the fourth day, retrofire was initiated over Canton Island; 34.5 minutes later, the flotation collar was attached, and the two astronauts were choppered to the flight dock of the USS *Wasp*.

At the postrecovery news conference, there were two themes: first, a mutual exchange of congratulations among all individuals and organizations involved in the program, including the media; second, the lessons learned of value to the achievement of lunar landing. These lessons were outlined by Bob Gilruth in terms of how to maneuver with precision, to rendezvous, to dock, to work outside in a hard vacuum, and to recover with precise landings.

Deke Slayton pointed out that the transition to Apollo was already occurring. All crews had been transferred to the lunar program except the crew of Gemini 12. Chris Kraft stressed the operational skills that had been built up by both ground and flight crews and added that the Gemini flight-control teams were being phased into the ultimate lunar mission. Dr. Charles Berry, the astronauts' many-talented flight surgeon, said that proof that man can really operate in the space environment was one of Gemini's milestones.[10]

Gemini Farewell

A week later, on 23 November 1966, there was a Gemini 12 pilots' news conference, followed by a Gemini awards ceremony. Jim Webb and I made the presentations. There were many deserving of recognition, but none more so then George Mueller, who had revitalized the mission soon after his arrival, and Chuck Mathews, the day-to-day manager. Others included the two most recent astronauts, Jim Lovell and Buzz Aldrin, as well as William Bergen, president of the Martin Company; David Lewis, president of McDonnell Aircraft Corporation; and James S. McDonnell, McDonnell Aircraft Corporation's executive officer. Mr. McDonnell seldom missed a Mercury or Gemini launching. He

9. Ivan D. Ertel, "Gemini XI-Mission-High Altitude, Tethered Flight," in *Gemini Program* (Houston, TX: NASA, 1967).

10. Ivan D. Ertel, "Gemini XII-Flight and Gemini Program Summary," in *Gemini Program* (Houston, TX: NASA, 1967).

would jog on the beach in the early morning, and after every flight, he would announce over the loudspeaker at one of his factories, "This is Mac calling; you've done it again. You've achieved another great success. Congratulations!" Morale at McDonnell was high because of the caliber of its employees, the great interest of Mr. Mac, and the frequent visits of the participating astronauts. Many other industry and government leaders received special awards; recipients included Major General Vincent G. Huston for his significant contribution in directing the efforts of the Eastern Test Range of USAF, in providing critical launch and operations support, and in directing the total efforts of the DOD operational support for the Gemini Program. Others recognized were the senior officers of the companies manufacturing the Agena and the Titan. It was a great day. As we listened to the Gemini accomplishments, I said, "Don't forget that there is much hard work ahead before we achieve our national goal of preeminence in space." Little did I know what lay ahead before a successful lunar landing and a safe return.

The rapid pace of manned test and flight missions abated somewhat following the completion of the Gemini flights and the program's awards ceremony. Reflection on the Apollo Program at this time would have shown progress in many quarters, including the Saturn I, Saturn IB, the F-1 engine, and the unmanned Surveyor and Lunar Orbiter programs.

Saturn I

The 10 Saturn I development flights started during President Kennedy's administration were completed in 1965. In September 1964, a boilerplate capsule, service module, and second stage elements weighing 37,000 pounds were placed in Earth orbit. The final three Saturns were used to launch Pegasus, a large spacecraft used to measure the size and number of micrometeorites in near-Earth orbit.[11]

Saturn IB

The second stage of Saturn I, with its six 15,000-pound RL10 engines, was replaced by the SIVb stage, which had a single J-2 liquid-hydrogen-fueled engine with a thrust of 200,000 pounds. The Saturn IB, as it was designated, was launched successfully twice in 1966. In August of 1966, 55,000 pounds of material was placed in Earth orbit.[12]

The F-1 Engine

The F-1 engine development that started in the Eisenhower era was an act of faith. Its 1.5 million pounds of thrust was an order of magnitude greater than the thrust of the existing engines used in ballistic missiles. The fuel and oxidizer pumps were driven by 55,000-horsepower engines, and the engine itself produced 160 million horsepower at ignition. The F-1 was a brute, and it was nearly impossible to tame. Blowups occurred, apparently at random, during its 3.5-minute burn. Its development was more an art than a science. The work was frustrating and didn't lend itself to mathematical analysis. In order to achieve consistency in the testing, bomblets were developed to upset the burning pattern. Then, if the engine was stable, the burning would recover in milliseconds. With this tool, holes in the injection plate for the oxidizer and fuel could be rearranged and quickly tested. Baffles of various dimensions could be introduced to determine their effectiveness. Finally, stability was achieved by moving the burn closer to the mouth of the nozzle, resulting in a loss in efficiency of only a few percent. The F-1 passed its flight-rating test on 8 March 1965.[13]

Saturn V

Of course, each of these giant engines had to function in close proximity to its four neighbors. A first-stage, full-duration test of the Saturn V was first achieved on a stand at Marshall Space Flight Center on 5 August 1965.

Similarly, tests were being run on the Saturn V second stage with its five J-2 liquid-oxygen, liquid-hydrogen engines. There were difficulties with the embrittlement of the hydrogen tank at its very low, cryogenically cooled temperature. In-house material scientists at Marshall resolved the issue, and the first successful full-duration firing occurred at the North American flight test stand on 9 August 1965.

---

11. Von Braun and Ordway, *History of Rocketry and Space Travel*, p. 167.

12. Ibid., p. 172.

13. Murray and Cox, *Apollo: The Race to the Moon*, pp. 144–151.

Figure 16. An aerial view of the Launch Complex 39 area shows the Vehicle Assembly Building (center), with the Launch Control Center on its right. On the west side (lower end) are (left to right) the Orbiter Processing Facility, Process Control Center, and Operations Support Building. To the east (upper end) are Launchpads 39A (right) and 39B (just above the VAB). The crawlerway stretches between the VAB and the launchpads toward the Atlantic Ocean, seen beyond them. (NASA Image 99PP-1213, also available at http://grin.hq.nasa.gov/ABSTRACTS/GPN-2000-000855.html)

The third stage of Saturn V was, by plan, similar to the second stage of the Saturn IB. However, for the lunar mission, this stage had to be ignited to complete the insertion into Earth orbit and then reignited for the extra thrust required for lunar passage. On 20 August 1965, the J-2 engine was ignited for 3 minutes, and then, after a 30-minute shutdown, it was reignited for the 4-minute burn that would later take the astronauts away from Earth and toward the Moon.[14]

In November 1963, when President Kennedy inspected Merritt Island from his helicopter, there wasn't much to see. The 4,800 pilings were still being driven through the sand to the bedrock below. Less than three years later, the Vertical Assembly Building was complete (see figure 16) and Saturn V was being fabricated. The first rollout of a full-fledged Saturn V mounted on the crawler transporter took place at a formal ceremony on 25 May 1966. At the appointed time, the doors opened, and the tremendous assemblage of hardware traveled forward at 5 mph. Apollo/Saturn V on the move (figure 17) was a deeply moving sight. With the 52-story VAB in the background, the crawler delivered the Saturn V to the launchpad erect, standing two-thirds the height of the Washington Monument. The congressional delegation, the guests, the press, and the NASA team were dwarfed physically and emotionally by such a majestic creation. The trip to the Moon was becoming a reality.

---

14. *Astronautics and Aeronautics, 1965: Chronology of Science, Technology, and Policy* (Washington, DC: NASA SP-4006, 1966).

Figure 17. The Apollo Saturn V 500F Facilities Test Vehicle, after conducting the VAB stacking operations, rolls out of the VAB on its way to Pad 39A to perform crawler, Launch Umbilical Tower, and pad operations. (NASA Image Number 67-H-1187.)

Surveyor—Unmanned Lunar Landing

As Jerry Wiesner said at the special meeting with President Kennedy in November 1962, "Jim [Webb] may not understand this, but we don't know a damn thing about the surface of the Moon." The point was well taken even though Jim did know. Tommy Gold, a professor at Cornell University, had a theory that the continuous barrage of meteors and micrometeors on the lunar surface had created not only voluminous craters, but also fine dust, into which a lunar lander might sink out of sight. If the truth be known, there was much we didn't know—and that was a good reason for exploration. However, to land there, we had to know something about the bearing strength of the surface.

Keith Glennan's last official act at NASA was to select Hughes Aircraft for the development of Surveyor. Initially conceived for unmanned exploration, the craft had become essential to accomplishing the lunar landing. But progress at Hughes was slow and a matter of deepening concern. It was decided that I should bait the bear and visit Pat Hyland, Hughes' chief executive officer.

In early 1966, I invited him to breakfast at a hotel near the Los Angeles International Airport. I came armed with two alternatives. One was a contract amendment that provided an incentive fee for Hughes. If they achieved a successful lunar landing prior to a given date, there was a bonus, and if there was a delay, there was a penalty that was increasingly stiff as the weeks increased. I also had a letter that laid out, in detail, specific errors in omission and commission by Hughes in the management of the Surveyor program. After pleasantries and a plate of scrambled eggs, I showed Pat the letter and the contract amendment and asked him which he'd like to receive (or like least to receive). He said he'd be happy to sign the contract document. I was at Mission Control in Houston for the launching of Gemini 9.

When the Agena failed, Gemini 9's launch was scrubbed because Agena would not be available for rendezvous and docking. So I headed for JPL in Pasadena, California.

In the early morning (2:00 a.m.) of 2 June 1966, I was seated on the balcony of Mission Control at the Jet Propulsion Laboratory, anticipating the landing of Surveyor. Pat Hyland was several rows behind. The atmosphere was palpable. Surveyor appeared healthy, responding correctly to instructions. Finally it made its landing, to great cheering; then it took the first photograph from the Moon, inspiring protracted cheering. And I heard Pat say over the din, "How's that for a crippled program?" And at last, we had the answer from the returning photographs. There was dust on the lunar surface, perhaps an inch deep. It appeared that the lunar surface would support a manned lunar landing.

Surveyor's 850-pound weight, was lifted into Earth's orbit by an Atlas-Agena. There were two more lunar landings, each in a designated area. The data from Surveyor were essential to the design of the Apollo lander, challenging to geophysicists, and awe-inspiring to the public.[15]

Lunar Orbiter for Mapping the Moon

The Fleming report of 1961 called for five Surveyor A landers and a series of Surveyor B orbiters, later called Lunar Orbiters. The objective of the latter was nearly complete mapping of the Moon and the location of landing sites. JPL was responsible for unmanned lunar and planetary missions, and hence, it was assumed, this laboratory would be in charge of both Surveyors A and B. However, its plate was full, and, somewhat related, NASA was experiencing difficulty renegotiating the contract with the California Institute of Technology, Caltech for short. JPL was and is an integral arm of Caltech. Both are located in Pasadena, California. It's not uncommon for the government to contract with a university for the management of a laboratory. The former AEC, now part of the Department of Energy, has a contract with the University of California for managing the Los Alamos and Lawrence Livermore National Laboratories, as well as one with the University of Chicago for managing Argonne. This type of arrangement provides more flexibility in personnel management than the civil service and can provide intellectual stimulus from university faculty.

15. *Astronautics and Aeronautics, 1966: Chronology of Science, Technology, and Policy* (Washington, DC: NASA SP-4007, 1967), pp. 203–205.

However, the monitoring of large contracts with the laboratory is more difficult than with a government center, and the fee paid to the university can be excessive, or appear to be, when an agency is obtaining congressional approval. JPL is a Federally Funded Research and Development Center (FFRDC). Further discussion of this type of management appears later in the book.

Since time was wasting, I reviewed the situation with Jim Webb and initiated discussions with Tommy Thompson, Director of NASA's Langley Research Center in Virginia. I asked them to investigate the transfer of the project to that Center's aegis. The technical team there was eager for the opportunity and appeared to have the necessary procurement and management skills. Langley then prepared a procurement plan, which I approved. The plan called for Langley to manage the project and team with a cross section of Langley and other NASA personnel. When Mr. Webb, Dr. Dryden, and I met with the evaluation team, the selection of contractors centered almost entirely on the technology, and a clear winner was not obvious. It was decided that I would hold hearings separately with each of the competitors. Hughes Aircraft proposed a rotating spacecraft similar to Syncom, its communication satellite. This design simplified the craft's stabilization but required higher sensitivity film. Boeing provided a stable platform aligned with the lunar vertical, so photographic requirements were less stringent than they might have been. In this design, each photograph was scanned into strips, lines, and finally dots. Each dot was digitized into a number on the gray scale from 0 (white) to 60 (black). The reconstituted photograph had more than adequate resolution. Since the main thrust of the mission was photographic, Boeing received the contract. The contract for five Lunar Orbiters was incentivized, with the contractor fee based on schedule and performance. Payments for the work had to be periodical. At one point, in order to avoid causing a renegotiation, NASA increased its payment to Boeing by $10 million. The reprogramming required congressional approval. Congress assented, the program held, and all five missions were successful. Both NASA and Boeing benefited from the incentive arrangement.

On 10 August 1966, Lunar Orbiter I was launched by an Atlas-Agena and headed for the Moon, where photographs of the lunar surface were obtained for 17 days. The high-resolution photography was disappointing, but the medium-resolution camera returned good images, though not of the quality we'd anticipated. The photos would cover a 3,000-mile strip around the equator while concentrating on nine potential landing areas. On 6 November, Lunar Orbiter II was launched; it provided 20 days of consistently high-quality photographs, including pictures of 13 potential landing sites.[16]

Having viewed a fair number of overhead crater photos, I asked how difficult it would be to obtain a few oblique shots of the lunar surface. The answer was "No problem; what would you like?" I suggested an interesting moonscape at an angle 10 degrees below the horizon. A few days later, I couldn't believe the result: on my desk was an image of the crater Copernicus, one of the great pictures of the 20th century. The photograph showed a pile of debris 1,000 feet tall in the center of the crater, with a flat, pockmarked surface around the debris and the 10,000-foot crater wall.

I was headed for Boston in a few hours to give a talk at the American Institute of Aeronautics and Astronautics' (AIAA) Third Annual Meeting and Technical Display. I asked for a slide of the picture and received it as I boarded the plane. When the photograph of Copernicus appeared in the auditorium later that afternoon, there was first stunned silence and then strenuous applause. What I said made little difference. I've had a copy of this photograph on my study wall ever since.

There were three more Lunar Orbiters, all successful. From the images taken by these orbiters, maps covering nearly 100 percent of the lunar surface were produced. As of year's end 1966, NASA was rapidly filling the squares on the PERT charts in the Fleming Committee's report of 16 June 1961. The committee that Fleming chaired studied every task that could be foreseen at the start of the manned lunar landing program. Each task was represented on the chart by a square. Many of these tasks were now complete, but much lay ahead.

---

16. *Astronautics and Aeronautics, 1966: Chronology of Science, Technology, and Policy* (Washington, DC: NASA SP-4007, 1967), pp. 262–265.

Organizational Issues

Toward the end of 1966, Willis Shapley and I had several command performances with Jim. He wasn't satisfied with NASA's organizational arrangements. It wasn't clear to us exactly why. In part, he felt that too much hinged on himself and me. Stated another way, he wanted to establish an organization that would outlast us and would carry NASA solidly into the future. I was concerned that radical changes would take time to accommodate, just as we were reaching the end of the decade, the time when we had hoped to achieve our lunar landing goal.

Jim Webb liked to receive information from a variety of sources. He sometimes called this "self-policing." He did rely heavily on Hugh Dryden and me, but he also surrounded himself with consultants of high caliber, and he often invited senior corporate officers for a discussion of the world scene. There was one occasion when Mervin Kelley, one of Jim's consultants, was convinced that the Draper Laboratory could not develop the guidance and control for Apollo. He felt that the laboratory was too theoretical and could not design hardware that would withstand the rigors of lunar missions. Fortunately, I was brought into the discussion and was able to enumerate many of the Draper Lab's successful developments.

I was not always enamored with some of Jim's creations, such as a secretariat. Documents emanating from both the Field Centers and Headquarters were categorized by the secretariat and distributed accordingly. I didn't believe that clerks understood the nuances of NASA's programs sufficiently to make such distributions. Also, metaphorically speaking, the turn of a key could isolate individuals from parts of a program without their knowledge, as happened to me later in the program.

A contract with General Electric (GE) also generated questions of intent. In February of 1962, NASA announced that GE had been selected for a supporting role in Apollo to provide safety analysis of the total space vehicle and to develop and operate a checkout system for Launch Control. Jim felt that this GE assignment would provide greater visibility of NASA's progress.

In actuality, the contract gave GE a major role in designing the hardware and software for Launch Control at the Cape, but their safety effort was secondary. GE had technical staff at Grumman, at North American, and in Houston. Their job was to look for designs and flight hardware that might create hazards for the astronauts. At North American, the GE engineers were often labeled as spies and were provided with trailers far from the Apollo effort. It was a difficult assignment for GE and not very productive.

Jim left for Princeton in December to give a series of lectures on modern management. He had a lot to say there about the Triad's advantages, one of his favorite hallmarks. When Jim returned, I invited him to lunch for a discussion of his and my views on organization. By dessert, he said that he needed my views in writing. My draft response dated 15 December 1966 outlined in detail what I had presented at lunch. I recommended a new Office of Management for all functions related to NASA's resources, i.e., financial operations, facilities, and manpower.

Harold B. Finger started his career at NACA's Lewis Research Center as an aeronautical research scientist in 1944. When NASA was formed, he was transferred to NASA Headquarters to become the chief of the Nuclear Engine Program. During the late fifties and early sixties, NASA and the Atomic Energy Commission (AEC) established joint projects for both engine and electrical power development. Harry was responsible for all these activities, which necessitated his wearing several hats. He was the manager of the joint AEC-NASA Nuclear Propulsion Office formed in 1960; he also served as NASA's Director of Nuclear Systems, and in June 1965, he became the head of a new AEC division for the development of electrical power for space vehicles. The isotope electrical power units were especially important to NASA on missions to the outer planets, at great distances from the Sun, where solar power became less effective. Harry had many constituents, and he satisfied them all. By 6 January 1961, Harry was asked to head a NASA team for the analysis of procedural revisions and functional alignments within NASA Headquarters and to review, in more depth, the specific options I had suggested in my discussions with Jim Webb. A more detailed discussion on this appears in the next chapter on NASA management.

The Apollo Fire

Instead of the glorious day that was anticipated, 27 January 1967 was a tragic day for NASA. Mr. Webb had invited the senior executives from both the Gemini Program and the Apollo Program to Washington to attend a White House ceremony, as well as a special dinner for the group.

The White House affair started formally with the signing of an international treaty. Representatives from 62 nations were involved in London, Moscow, and Washington, DC. Secretary of State Dean Rusk signed for the United States in the presence of Soviet Ambassador Anatoly Dobrynin, British Ambassador Dean, U.S. Ambassador to the United Nations (UN) Arthur Goldberg, and President Johnson. The President described the treaty as the "first firm step toward keeping outer space free for ever from the implements of war."[17] After an exchange of pleasantries with our Apollo executives, I headed for home, where I was having a small dinner in honor of my old boss from MIT, Doc Draper. As I walked in the front door, I heard my wife Gene say, "Here comes Bobby now." It was George Low on the phone, and his first words were, "They're all dead." He was extremely upset, and I was having difficulty in understanding him. I asked George, "Who's dead?" Gradually, or so it seemed, I learned that a fire had started during the testing of Apollo 204 (later known as Apollo 1) and that the three astronauts on board were dead. Those three were Gus Grissom, of Mercury and Gemini fame; Ed White, the first U.S. astronaut to experience extravehicular activities; and Roger Chaffee, gaining experience for what would have been his first spaceflight. As I absorbed the devastating news, I realized that I must immediately leave for my office, and I told George that I would call him in a half hour for a more complete report. I then asked Gene to take over as host of the dinner and tell the guests of the accident only as they were departing.

Back in my office, I soon learned that a full-scale test of the Apollo system had been under way when the fire started. Not only were the astronauts enclosed in the capsule in communication with Launch Control, but the worldwide Apollo net was also involved. Once the fire started, the pressure inside the capsule increased, so the hatch could not be opened. Ultimately, the capsule burst and the flames scorched the outside, but fortunately, the fire did not spread to the surrounding structure.

My first job was to communicate with Jim Webb and George Mueller and to ensure that all senior individuals were notified. While I was talking to Bob McNamara's assistant, the operator cut in with an emergency call; it was Peter Hackes of the National Broadcasting Company (NBC). He asked me to come immediately to the studio to explain the fatal fire. He said that the country was in near panic. I explained that I couldn't, that I was busy making important arrangements, and that the cause of the fire was still unknown. Jim Webb was busy gaining acceptance from the President and Congress for a NASA investigation. George Mueller was gathering detailed information about the fire and preparing a list of individuals to serve on the accident review board. Following the Gemini 8 near-accident with Armstrong and Scott aboard, I had revised the manual to be followed when accidents occurred. NASA had dealt with fatal aircraft accidents in the previous five years, but none involving space activities. I drew up and signed the instructions for the review board and caught a few hours' sleep before proceeding to Langley Field early the next morning to pick up Tommy Thompson, Director of the Langley Research Center. We then headed for Cape Canaveral, where he was to become the chairman of the review board.

As soon as we arrived, we met with Apollo Program Manager Sam Phillips, Joe Shea, and Kurt Debus (the Director of Kennedy Space Center). I asked for a brief status report and then brought them up to date on the plans for the accident review board. I advised them that all hardware and software utilized in the Apollo 204 test had been impounded, and its release for inspection and testing was the sole responsibility of the board. I provided them with a list of the board members and told them that we were attempting to shield the board from direct inquiry by the media. While I was there, the media requested an interview directly with me. I notified the

---

17. *Astronautics and Aeronautics, 1967: Chronology of Science, Technology, and Policy* (Washington, DC: NASA SP-4008, 1968), p. 23.

TV and press that time didn't permit a press conference. The release made at the time of my departure stated that a board had been established to review the circumstances surrounding the accident, to establish probable cause, and to review the corrective action and recommendations being developed by the program office, Field Centers, and contractors involved. The board was required to document its findings, determinations, and recommendations and to submit a final report to the Administrator, not to be released without his approval.

All three astronauts were buried with full military honors—Gus Grissom and Roger Chaffee at Arlington Cemetery, and Ed White at West Point. The President, Jim Webb, and my wife Gene were at Arlington, along with many others; I accompanied Mrs. Johnson and the Vice President to the service in the chapel at West Point. I met Ed's father, told him how much his son contributed to the space effort, and apologized for allowing the fire to take place.

The President, Congress, and the media wouldn't sit still for an extended period without any information on the board's progress. The barely acceptable solution required me to visit with the board weekly, listen to their progress report, and then draw my own conclusions. While flying back to Washington, I'd write my own findings and submit the result to Jim Webb. If acceptable to him, the report would be transmitted in sequence to the President, then to the House and Senate space committees, and, ultimately, to the media.

My Reports on the Progress of the Accident Review Board

My first report, dated 3 February 1967, stated that full advantage was being taken of the extensive taped data from the test, as well as records made prior to the accident. The report noted that the spacecraft was still mated to the unfueled launch vehicle at the pad. The report went on, "The capsule will be disassembled so that experts in many technical and scientific areas can work with the physical evidence, and an undamaged and nearly identical spacecraft will be used to establish the condition prior to the accident."[18]

The report contained a timeline of the events following the crew's detection of the fire. At 6:31:03, pilot Chaffee reported that a fire existed. One second later, the inertial navigation gave an indication of crew movement. The cabin temperature began to rise after 2 seconds, and senior pilot White reported the fire after 6 seconds. At the same time, the pressure started to increase and a large amount of astronaut motion was detected. Nine seconds after the first indication, pilot Chaffee reported a bad fire. There was no further intelligible communication. After 14 seconds, the pressure and temperature of the astronauts' suits commenced to fluctuate and the signal was lost. Soon thereafter, the pressure in the cabin doubled and the capsule skin ruptured. The cause of death was asphyxiation due to smoke inhalation. My first report was printed verbatim in the *New York Times*, five days after my trip to the Cape.[19]

My second report, dated 14 February, commented on the board's structure and procedures. "Approximately 5,000 scientists, engineers and technicians are involved in the investigation. 21 panels have been established to conduct the inquiry. No single element is touched or removed for analysis without full board approval to ensure there is no impact on the on-going studies. All three space suits were burned through and Gus Grissom received the greatest exposure. The cause of the fire has not been determined."[20]

Prior to the completion of the board's report, the press was relatively kind to NASA. *Business Week* stated, "No previous frontier has ever been crossed without loss of life. It was not to be expected that space, the most perilous frontier of them all could be conquered without sacrifice." The *Washington Evening Star* found that "second guessers are wondering whether we should be going to the moon at all. From any rational point of view, the only thing to do is carry on." However, *Los Angeles Times* editor Marvin Miles accused NASA of "shortsightedness and trying to hide its negligence." But *Technology Week* responded, "Its [sic] our impression the agency is trying valiantly to come up with just such information (why the hatch could not be opened)."[21]

18. Robert C. Seamans, Jr., papers, MC 247, Institute Archives and Special Collections, MIT Libraries, Cambridge, MA, and *New York Times* (5 February 1967): B13.

19. Ibid.

20. Ibid.

21. *Astronautics and Aeronautics, 1967: Chronology of Science, Technology, and Policy* (Washington, DC: NASA SP-4008, 1968), pp. 57, 64–71.

I prepared the third and last report for Mr. Webb based on my visit to the Cape on 25 February. The board expressed the view that the experience in tests and in flights prior to the accident suggested that the probability of fire was low. Neither the crew nor the development and test personnel considered the risk of fire to be high. The board did not recommend changing the pure-oxygen system or the planned cabin pressure.[22]

Monday, 27 February 1967, Senate Hearing

In testimony before the Senate Committee on Aeronautical and Space Science on 27 February, George Mueller outlined an extensive program of redesign and testing, as well as a number of procedural changes. Specifics included an escape hatch that could be opened in 2 seconds, a search for new and less flammable materials for the cabin and spacesuits, and a review of emergency procedures. He guaranteed that all improvements and changes would be incorporated into an advanced version of the Block II Apollo spacecraft. During the question period, Senator Mondale asked about a report on the performance of North American Aviation. Mr. Webb referred the question to George Mueller and Sam Phillips in turn. Neither one had knowledge of such a report, although a NASA tiger team had investigated North American and had found faulty workmanship, spotty organization, and other deficiencies. Sam had reported the results to me at the previous December status review. I wondered to myself if Senator Mondale had seen the findings and recommendations of our study. So I explained that from time to time, NASA had onsite reviews of contractor progress and that the information Senator Mondale was referencing might be in this category. When the session was over, Jim Webb told me to return to Headquarters with him and our General Counsel, Paul Dembling. Jim's "limo" was a Checker cab painted black with a window between the driver and the backseat occupants. Once aboard, Jim quickly cranked up the window and lacerated me in no uncertain terms. These hearings weren't the love-fests we normally had had with Congress in the past. Millions of dollars could be riding on the outcome, and under no circum-stances should information be volunteered. I said I believed that Senator Mondale was using a set of transparencies used by the tiger team in their presentation. Jim cut me off almost before I'd finished the sentence. The lecture continued until we left the car. As I was recovering my equilibrium, Paul came into my office holding a thick document. The first two words on the first page were "This report." Sure enough, the report was a bound copy of all the transparencies used by the tiger team when it reported on its review of North American. The first page merely explained the circumstances for the investigation. I should have taken this document into Jim, but my instant reaction was for Paul to do the honors so that I could sit in my office, catch my breath, and review the situation.

World Travel

I'd planned an extensive NASA trip around the world prior to the Apollo fire. Was it appropriate to be traveling around the world at this time? Jim Webb felt that Gene and I should still go. So the first stop was Paris to attend the Advisory Group for Aerospace Research and Development (AGARD), a North Atlantic Treaty Organisation (NATO) scientific and technical meeting. Then it was off to Kenya, with a plane change in Rome, where we were joined by Professor Luigi Brolio, who headed the San Marco project, a joint U.S.-Italian project for launching a satellite from the Indian Ocean near the equator. The United States was providing the boosters; the Italians, everything else, including the platforms and the satellite. When we arrived at the airport in Rome, it was a mass of cables. Joseph Stalin's daughter had just asked for asylum in the United States; she had left India and was on her way to New York. Later in the day, we changed planes in Nairobi for Mombasa on the coast. It was 11 March, and we celebrated Gene's birthday there. The next day, an Arab drove us in a Jeep along the coast to Campa Basa, the Italian base camp. Rubber boats took us to the two Italian platforms in the Indian Ocean, where we inspected their preparation for a satellite launch and enjoyed a delicious lunch outdoors near the equator. The next stop, in Bombay, India, provided me with an opportunity to visit the Tata Research Institute, a rocket-development site, and a nuclear facility. We

---

22. Ibid.

then had a lovely day in New Delhi with old friends Galen and Ann Stone. He was in charge of our embassy when Stalin's daughter knocked on the door. On 17 March, we arrived at Honeysuckle Creek Tracking Station near Canberra, Australia. The then-newest member of NASA's 16-station network for Apollo missions was dedicated by the Australian Prime Minister, Henry Holt, and me, with an assist from Ed Buckley, the NASA Associate Administrator for Tracking and Data Acquisition. Vice President Humphrey cabled his congratulations to the Prime Minister, and then the two had an animated conversation.

Saturday, 15 April 1967, Final Report on Apollo 204 Fire

The Apollo 204 review board final report was submitted to Mr. Webb on 5 April 1967. The board identified the conditions that had led to the disaster as follows:

1. A sealed cabin, pressurized with a high-pressure oxygen atmosphere

2. Extensive combustible material in the cabin

3. Vulnerable wiring

4. Inadequate provision for escape or rescue

This report provided 21 recommendations, including the following:

1. Review of life-support system

2. Investigation of effective ways to control and extinguish cabin fire

3. Severe restrictions on combustible material

4. Reduction in time required for astronauts to egress in emergency

5. Continued study of two-gas cabin atmosphere

6. Full-scale mock-up tests[23]

The accident review board, under Tommy Thompson, with astronaut Frank Borman as the spokesman, performed a wonderful service for NASA. Their conclusions and recommendations were sound and inclusive. Frank was articulate as he presented the information and answered questions. In particular, he said that if the findings were followed, he would have no problem stepping into the capsule himself.

From my own standpoint, I didn't feel that NASA and its contractors required major surgery. Obviously management and procedures can be improved, but faulty administration didn't cause the loss of Apollo 204. Rather, it was an error in engineering judgment, and we were all guilty. Astronauts should never have been subjected to 14.7 pounds per square inch, or psi (sea-level pressure) of pure oxygen. Once a fire starts under this condition, it cannot be suppressed. Before the fire, NASA tested all aspects of the equipment to be used in flight. There were tests for vibration, temperature, and pressure. Rocket motors were fired many times, as were each of the Saturn I and Saturn V stages. But NASA never tested a boilerplate capsule for fire. We would have been horrified by the result, the fire would have been so intense. However, at the partial pressure of oxygen as we find it in the atmosphere at sea level (3.5 psi), the burn rate is the same for a single gas as for multiple gases, as found in nature.

A single-gas system was selected for good reasons. Number one was simplicity. Only one system of tanks and controls was required to feed oxygen into the capsule and the astronauts' pressurized suits. Also, there was no concern about rapid pressure changes of nitrogen, which can lead to physiological problems including the bends. So pure oxygen was used with great success in both Mercury and Gemini capsules, and on the basis of this experience, NASA continued the same practice in Apollo. The mistake was not in the use of pure oxygen, but in filling the capsule and astronauts with pure oxygen at sea-level pressure. If the oxygen had been maintained at 3.5 psi while the nitrogen had been bled out as the Apollo went into orbit, fire could have been contained and extinguished.

---

23. *Astronautics and Aeronautics, 1967: Chronology of Science, Technology, and Policy* (Washington, DC: NASA SP-4008, 1968), p. 23.

North American Aviation (NAA) fell from grace as a result of the fire. In our congressional hearings, it became known that a NASA tiger team had censured them for sloppy workmanship. They countered with their early recommendation for a two-gas system. It's true that North American had recommended a two-gas system; however, NASA maintained its single-gas position for the reasons mentioned in the previous paragraphs. NAA was selected as the contractor for the Apollo capsule and service module even though Martin Marietta was scored higher by the source evaluation board. At the meeting after the board's presentation, Bob Gilruth met with Jim Webb and the other members of the Triad. He expressed concern that Martin hadn't had any aircraft experience for years and Apollo was to be flown by the astronauts. We then listed recent NAA aircraft experience. There were many, with the X-15 heading the list. The X-15 had had many successful flights, both inside and outside the atmosphere, at speeds up to 7,000 mph. After Bob left the room, we examined Martin's strengths and determined that they excelled in areas not as key to the success of Apollo as North American's high-speed flight capability. Hugh Dryden prepared a handwritten list of all aircraft developed and produced by North American. I kept his note in my files, and it became a most useful document during our congressional hearings when the reason for our selection was coming into question. NASA was exonerated from selection complicity, but our ability to manage was still in question and had to be proven again.

Sunday, 23 April 1967, Soyuz I, Vladimir Komarov's Fatal Flight

On 23 April, the USSR launched its Soyuz I spacecraft with a single cosmonaut, Vladimir Komarov, in control. After completing his mission, Komarov attempted to reenter the atmosphere, but failed when he was unable to control his spacecraft. On the 18th orbit, after successful braking for reentry, the parachute lines apparently became snarled and the "spacecraft descended at great speed." Komarov was buried in the Kremlin Wall. President Johnson, Vice President Humphrey, Mr. Webb, and our astronauts all sent messages of sympathy to the Soviet Union.[24]

Tuesday, 9 May 1967, Senate Hearing Program and Contractual Changes

Jim Webb, George Mueller, and I appeared before the Senate Committee on Aeronautical and Space Sciences on 9 May 1967. Mr. Webb noted that arrangements for some of the prime Apollo companies were being realigned. NASA had negotiated a strong incentive contract with North American for the fabrication, testing, and delivery of Block II spacecraft and had expanded Boeing's responsibilities to include integration of all elements in the Apollo/Saturn stack. These included the three stages of Saturn V, the instrument unit, the Lunar Lander, and the Apollo capsule and service module. I commented on the schedule, saying that landing before 1970 remained possible. George Mueller discussed in detail NASA's response to the Apollo review board. For the time being at least, George testified that a single-gas system would continue to be utilized. All other recommendations of the board would be followed, and in addition, a new Office of Flight Safety was being established to evaluate safety provisions and monitor test operations. The officer in charge would report directly to George.

Thursday, 9 November 1967, First Flight of Saturn V

Thursday, 9 November, was the day planned to determine in dramatic fashion the validity of all-up systems testing. Four flights of Saturn I were allocated to tests of its first stage before the second stage was included. All four flights were completely successful, and the option to move Saturn I's development ahead faster was delayed two years. How different the first Saturn V flight test was. On the launchpad were the three stages of Saturn V, the Saturn instrument package, the Apollo capsule, and its service module. There was the whole enchilada except the lunar excursion module (LEM). A press conference was held outdoors the day before the launch, with Kurt Debus and me officiating. We faced over 1,000 members of the media with the Saturn V steaming behind us. The remarkable backdrop was awesome. At the press conference, I explained that in addition to testing all three stages of the Saturn V, we were using the service module rocket engine to take

---

24. *Astronautics and Aeronautics, 1967: Chronology of Science, Technology, and Policy* (Washington, DC: NASA SP-4008, 1968), pp. 101–102.

Figure 18. On 9 November 1967, Apollo 4, the test flight of the Apollo/Saturn V space vehicle, was launched from Kennedy Space Center Launch Complex 39. This was an unmanned test flight intended to prove that the complex Saturn V rocket could perform its requirements. (NASA Image Number 67-60629, also available at http://grin.hq.nasa.gov/ABSTRACTS/GPN-2000-000044.html)

the capsule to a higher altitude and drive it back into the atmosphere at near-lunar-reentry velocity. With such a large press corps, including Soviet and other foreign correspondents, there were bound to be tough questions—and there were. Weren't we taking too big a risk with all-up testing? If the first stage exploded, could the astronauts escape? Afterwards, I apologized to Kurt Debus for the press conference ordeal. Don't, he said; a free press is essential to democracy. He added that during World War II in Nazi Germany, he had had no information on the war's progress except what was spoon-fed by Joseph Goebbels. Kurt continued, "I really believed the news Goebbels propagated."

The countdown to the Saturn's launch proceeded without a hitch; all seated in the viewing stands heard the tremendous pent-up energy suddenly being released when the Saturn V reached the position shown in figure 18. Saturn V was just nearly clearing the tower, and the sound was just reaching the viewing stand. The sight and sound were truly awesome. The sound was heard by the ear as lots of noise crackling and popping, and by the body as a rumbling vibration. Dr. William Donn of Columbia University found the Saturn V blastoff to be one of the loudest natural or manmade noises in history, excepting nuclear detonations.

Early indications from the first Saturn V flight (also called the Apollo 4 mission) were all favorable. Later, analyses of the data showed that the thrusts of all the engines were well within tolerances and that the capsule approached Earth's atmosphere at close to the nominal 7 degrees below the horizon and at a speed of 24,900 miles per hour. [25]

The Command Module landed near Hawaii and was picked up successfully by the USS *Bennington* 2.25 hours after splashdown. The rocket motors, the structure, the controls, the instrumentation, the guidance, and the heatshield all had been completely successful. The members of NASA's highly professional rocket team, headed by Wernher von Braun, were astounded, and George Mueller was vindicated for his bold planning and execution.

Sam Phillips was moved to say that he was tremendously impressed with the smooth teamwork exhibited. Werner von Braun said that no single

event since the formation of the Marshall Space Flight Center in 1960 equaled that day's launch in significance. Jim Webb praised the devotion and quality workmanship of the 300,000 men and women working on the Apollo Program. And President Johnson said, "The whole world could see the awesome sight of the first launch of what is now the largest rocket ever flown. This launching symbolizes the power this nation is harnessing for the peaceful exploration of space."

As Jim Webb said, well over a quarter of a million individuals were responsible for the Apollo mission and the flights to follow. And within this team were many leaders from universities, industry, and the government. Those in NASA with major management responsibilities who deserve great credit obviously include the following:

- Keith Glennan, who formed NASA and pushed it hard

- Jim Webb, who kept an umbrella over our heads even in stormy times

- Hugh Dryden, a respected scientist who understood the machinations of government

- George Mueller, who brought new ideas and experienced personnel to NASA with steely-eyed precision

- Joe Shea, who shifted John Houbolt's lunar orbit rendezvous onto the front burner and then managed the spacecraft development

- George Low, who started early in the program and stayed late, holding many key assignments

- Sam Phillips, former project leader of the Minuteman ICBM, whose experience in juggling many balls was essential to success

- Abe Silverstein, who helped get Apollo off to a fast start and provided assistance in his director's role at Lewis Research Center

- Tommy Thompson, who spawned the Space Task Group and chaired the Apollo accident review board

---

24. *Astronautics and Aeronautics, 1967: Chronology of Science, Technology, and Policy* (Washington, DC: NASA SP-4008, 1968), pp. 339–341.

- Eberhard Ress, the general manager who made Wernher von Braun's visions come true

- Rocco Petrone, who was responsible for the zero stage of Apollo, the massive ground facilities at Kennedy Space Center

- Edmund Buckley, who provided the necessary world communication network for tracking and data acquisition

And the leaders of the three Apollo Centers:

- Wernher von Braun, Director of Marshall Space Flight Center

- Bob Gilruth, Director of the Manned Spacecraft Center

- Kurt Debus, Director of John F. Kennedy Space Center

All of these people had to deal with many difficult issues.

This is an important but incomplete list of those who had major management responsibilities for manned spaceflight in the 1960s. There was also a large cadre of scientists and engineers who jumped into the breach on many occasions. I especially would like to mention Max Faget and John Houbolt. Max Faget's spacecraft designs, with their blunt bodies forward, brought our astronauts successfully through flaming reentries and back to Earth. Without John's persistence and creativity, we would not have selected the lunar orbit rendezvous mode for the lunar landing and we would not have successfully landed on the Moon. I was extremely fortunate to work with such talented individuals. When I arrived at NASA, Mercury was front and center, and our objective, as President Eisenhower indicated at a Cabinet meeting, was to accomplish as much as possible for $1 billion. And Keith Glennan, the first Administrator, did an excellent job with these funds, laying the groundwork for what was to follow.

When the Soviets threw down the gauntlet for the fifth time with the Gagarin flight, President Kennedy accepted the challenge and NASA embarked on Apollo, a most ambitious program. As general manager of NASA for seven years, I had overall responsibility for all aspects of NASA research and project planning, development, and flight operations, both manned and unmanned. In this monograph, I have attempted to outline the steps that NASA took to advance manned spaceflight during my tenure. Important though the unmanned programs for science, meteorology, and communications were, I included in this monograph only those unmanned projects directly relevant to the Apollo landing. The following chapter describes NASA's organization and the tools that I used as general manager of NASA from 1 September 1960 to January 1968.

Chapter 5:
NASA MANAGEMENT

NASA was formed from a number of separate entities, and hence was a hybrid organization. Four of its Centers were formerly the action arms of the National Advisory Committee for Aeronautics, one was central to development at the Army's Redstone Arsenal, one grew from a Navy research team, and one was a nonprofit organization managed by the California Institute of Technology. A diagram of their antecedents is shown in figure 5 (see chapter 3), along with the Soviet counterpart. Each of these teams had a nucleus of highly qualified leaders supported by strong scientific and technical personnel.

Langley Research Center was the original NACA laboratory formed in 1915. Its pilotless aircraft division assumed responsibility for the Mercury program in NASA's first six months of operation. The so-called Space Task Group became 1,000 strong at Langley before transferring to Houston, Texas, to become the Manned Spacecraft Center, ultimately renamed Johnson Space Center.

Originally, flight operations were conducted at Cape Canaveral on a project-by-project basis, with the responsibilities vested in the Space Task Group for Mercury, Marshall Space Flight Center for Saturn, and JPL and Goddard Space Flight Center for the unmanned satellites and probes. The coordination of the resulting projects with the management of the Air Force Atlantic Missile Range became too unwieldy, especially when the Mercury program was joined by the addition of Gemini and then Apollo. So in 1961, the total Cape effort was integrated under a single director and became known as the Space Flight Center until later, when it became the John F. Kennedy Space Center.

NACA-Derived Centers

The four Centers derived from the NACA were the already-mentioned Langley Research Center in Hampton, Virginia, as well as Ames Research Center in Moffett Field, California; Lewis Research Center in Cleveland, Ohio (now called Glenn Research Center); and Dryden Flight Research Center in Edwards, California. Langley was more general-purpose and project-oriented, Ames concentrated on supersonic flight, Lewis concentrated on propulsion, and Dryden was for flight-testing.

At the time NASA was formed, the NACA was devoted almost entirely to aeronautics—astronautical studies were not encouraged. Its Centers were strictly for research and testing, with a wide variety of supporting wind tunnels and other test facilities. These Centers were recognized as preeminent in their fields by both the military and industry. Top-grade personnel were attracted to, and retained by, these Centers because of the importance of their research and the second-to-none tools available for aeronautical studies. Their charter was to support military and commercial aviation. The decision to conduct specific studies was entirely the responsibility of the Centers and NACA management. Suggestions to conduct the efforts on a reimbursable basis were always quashed for fear of NACA's becoming a "job shop" and thereby losing control. However, both the military and industry provided wind tunnel models, test equipment, and, on many occasions, full-scale aircraft for research and test purposes. For example, the X-15 aircraft was financed by the Air Force in consultation with the NACA; it was designed and built by North American Aviation under contract to the Air Force; and the flight research was the responsibility of the NACA.

When NASA was formed, the role of the NACA Centers was expanded to include aerospace research and, in some cases, actual project responsibility—for example, the Mercury Space Task Group and the unmanned Lunar Orbiter, both at Langley, and the Agena launch vehicles at Lewis. The research funding for aeronautics was carried as a line item in the budget, as was some of the funding for aerospace studies. However, there was also a supporting research and technology subline item for each manned and unmanned space project. These funds were distributed throughout NASA. I remember Hugh Dryden's admonition to me: "Don't let them include the supporting research as a lump-sum line item, it's much too easy for the Congress to dissect and eliminate."

When technical problems arose in the management of major projects, it was most advantageous to have available research personnel at both the research and the flight Centers. This was the case in a major way after the Apollo fire; individuals truly knowledgeable and coming from a wide variety of fields could be immediately deployed to the accident review committee conducting this investigation. In times of crisis, a strictly project organization must reach out to other organizations, usually on contract. Then, if the investigating consulting firm's conclusion differs from that of the contractor that experienced the accident, whom does the government believe? When conducting advanced technical efforts, it's imperative to maintain in-house technical skills of a high order. But high-grade technical personnel cannot be stockpiled. They must be given real rabbits to chase or they will lose their cutting edge and eventually seek other employment.

The Centers for Unmanned Missions

Goddard Space Flight Center

The Goddard Space Flight Center was an off-shoot of the Vanguard Project that was managed by the Naval Research Laboratory. The Center had primary responsibility for geophysical and solar research, astronautical observatories, and applications such as meteorological and communication satellites. This Center also was responsible for the tracking and communication stations for near-Earth manned and unmanned vehicles. Harry Goett was its first Director. He and his boss, Abe Silverstein, Associate Administrator for Space Programs, attracted an excellent team that managed a wide variety of unmanned space vehicles and even, for a time, the Space Task Group at Langley.

Jet Propulsion Laboratory

The Jet Propulsion Laboratory operated under the aegis of the California Institute of Technology on contract with NASA. In the lingo of today, JPL comes under the rubric of Federally Funded Research and Development Center, or FFRDC. See chapter 4 for an earlier discussion of this type or organization.

JPL was responsible for all unmanned lunar and planetary vehicles, including the already-discussed Ranger and Surveyor programs, as well as Mariner and Voyager, in addition to the Deep Space Network. JPL had a great deal more flexibility in its personnel management than the other NASA Centers. Most other NASA employees were civil servants. However, NASA was allowed to hire a certain number of employees to fill excepted positions at the discretion of the Administrator. This permitted prompt hiring of exceptional candidates who could later be folded into government civil service positions at the appropriate level. And many of NASA's senior personnel, such as the Associate Administrators, continued as excepted employees throughout their tenure.

Two major issues festered amongst the NASA-Caltech-JPL threesome, one internal to JPL and the other primarily with Caltech. The contract with Caltech expired in 1963. Caltech was receiving a substantial fee, and for what purpose? Caltech's primary responsibility was selecting the JPL Director, certainly not a major budget drain. Lee Dubridge, president of Caltech, said that the fee compensated for the risk to his institution if there were major accidents. On the positive side, he said that JPL, by receiving its funds through Caltech, was given the aura of a premier institution. And, he added, hiring key personnel was facilitated by joint appointments at JPL and Caltech, although there were only a few of these. It was hard to justify the fee; as a result, Congress, and especially the chairman of our appropriation subcommittee, tried to restrain or eliminate funding for JPL. The impasse was partially resolved by transferring the responsibility for the Lunar Orbiter to Langley. This improved NASA's negotiating leverage with Caltech and resulted in a fee reduction. Then we hardballed the subcommittee chairman, Albert Thomas, by coupling the funding for the Manned Spacecraft Center in Houston (his district) with the funding for JPL.

Difficulties in the management of JPL were much easier to address. Bill Pickering, the Director, was a talented leader, but not strong on administration. For example, when the Ranger spacecraft was undergoing final vacuum tests, new components would be introduced as flight articles without authorization. Although the components might have improved performance, other parts of the spacecraft might have been adversely affected by their inclusion. After discussion and prodding, Al Luedecke became the manager of the Laboratory. He'd had similar responsibilities in the Atomic

Energy Commission. From then on, the discipline in the Laboratory was materially improved.

I'm afraid I was on JPL's most wanted list—that is, most wanted to leave—for pressing hard for organizational changes, but thanks to all concerned, JPL has conducted outstanding scientific investigations of our planets and their moons in the last quarter century.

The Centers for Manned Spaceflight

Marshall Space Flight Center

Marshall Space Flight Center has a strong heritage that dates back to World War II. Wernher von Braun worked at Peenemünde, the center on the Baltic where the Germans developed their "vengeance" weapons, the V-1 and the V-2. The V-2 was a ballistic missile that attacked targets at supersonic speed. Means didn't exist for defense against it, but fortunately, they weren't available until near the war's end. Wernher and his team managed to circumvent the Gestapo and were later captured by the United States. This group, which included many of the top leadership, fled to Bavaria, bringing with them reports and drawings. After their capture by the United States, they were whisked to the White Sands Proving Ground as prisoners of war. Over time, they became U.S. citizens, were joined by their families, and became managers of the research and development programs at the Army Redstone Arsenal. Prior to leaving White Sands, they converted captured V-2 weapons into sounding rockets. At the Redstone Arsenal, they developed newer, more advanced missiles. Their Redstone missile launched Alan Shepard and Gus Grissom into suborbital flight. Prior to that, they launched the JPL-built Explorer 1, the first U.S. satellite, into orbit. For a more detailed account of their exodus from Peenemünde, see pages 114 through 118 of *History of Rocketry and Space Travel* by Wernher von Braun and Frederick Ordway III.

Wernher and his team had trouble letting go of projects. Arsenal types by experience, they developed, constructed, tested, and launched vehicles with precision, but they had limited skill in contracting with industry for these capabilities. At one point, they recommended canceling the contract with General Dynamics for the Centaur booster. A Headquarters command decision immediately transferred the responsibility of Centaur to Lewis Research Center. Since then, the Centaur has become a reliable workhorse providing propulsion for many important missions.

Real progress was made when Bob Young from Aerojet Corporation joined Wernher and established a project office for each of the major projects under Marshall's control. Bob had a project director for the F-1 and J-2 rocket engines, each of the two stages of Saturn I, the first and second stages of Saturn V, and the instrument module. The third stage of Saturn V was similar to the second stage of Saturn I, and the project management was shared. A contract was responsible for the design, fabrication, and testing of each unit. Also, as previously mentioned, Boeing was responsible not only for the first stage of Saturn V, but also for integrating all three stages of the Saturn V vehicle and, in turn, the Apollo and the Lunar Lander.

Manned Spacecraft Center

The Manned Spacecraft Center, now the Lyndon Johnson Space Center, was transported from Tidewater, Virginia, to the lowlands of Houston. Water transportation for low-draft craft was available from the Center to Galveston Bay. At the time of acquisition, this attribute was included as a requirement, but its use was never exercised. At the time of President Kennedy's visit in December of 1962, the Center existed in rental space throughout Houston. However, by the time Gemini missions began, the Center was in full swing on acreage contributed by the city fathers.

Mission Control was the nexus of the Center. That's where the astronauts and mission hardware came together with the worldwide tracking and communication network. There were a variety of flight simulators that could be introduced to add realism to rendezvous and lunar landing operations. In addition to running regular missions, a wide assortment of adverse situations could be introduced to educate and test the skills of all participants.

This Center was home for the astronauts. Their training started here. Their medical testing and physical fitness programs were conducted at the Center, as were a wide variety of special simulators for docking with other craft and conducting extravehicular activities. This latter type of testing and training is most accurately simulated under water. Today, the Center has a pool large enough to test major sections of the International Space Station.

Kennedy Space Center

The launch facilities at the Cape, now incorporated into the John F. Kennedy Space Center, were responsible for what was often called the zero stage. This Center interfaced in many ways with the vehicles to be launched. The vehicles had to be mounted on the launchpad, either from the start, when under assembly, or, in the case of Apollo/Saturn V, after being assembled in the VAB and transported on the crawler to the pad. There were hold-down clamps used for 3 or 4 seconds to be certain that the appropriate thrust was obtained and swingback arms disconnecting electrical and hydraulic connections. Near the pad were liquid-hydrogen and -oxygen tanks to keep topping off the vehicle tanks that were continually evaporating and emitting gas into the atmosphere.

There was much coordination required with the Air Force's missile range. The most critical area was range safety. If a vehicle veered off toward a populated area such as Miami, it had to be immediately destroyed. But with the astronauts aboard, time was needed for their escape. With Mercury and Apollo, the escape rocket pulled the capsule clear of a potential accident, but with Gemini, there was a seat-ejection system—the capsule remained attached to the launch vehicle. In all cases, parachutes brought the astronauts back to Earth.

Lessons Learned at the Centers

NASA gained experience with two types of center organization: civil service centers and JPL, an FFRDC. In making a comparison between the two forms, it must be recognized that when government resources are used, the government is accountable for the expenditures. For this reason, functions such as procurement, launch, and flight operations should remain under direct NASA control.

As has already been mentioned, the use of excepted positions alleviated the difficulty of hiring key personnel, but many were later rolled into the civil service. However, cutbacks in civil service personnel were much more difficult. Firing an individual requires extensive liturgical-type proceedings with the individual, all under the oversight of the Civil Service Commission. A reduction in force (RIF) is more manageable, but in that instance, jobs are eliminated, not individuals. When a job is no longer required, the individual holding the job may have seniority over another individual and take over his or her billet. To defeat this difficulty, Marshall Space Flight Center is reported to have developed a special personnel computer program to release specific individuals. Their scheme may be apocryphal, but it was believable to those who have dealt with management issues in the government. The FFRDC provided much greater personnel flexibility, including the possibility of higher salaries for those carrying greater responsibility, although their salaries and total numbers do come under congressional scrutiny. However, FFRDCs have limitations. The government must take overall responsibility for procurement, including second- as well as first-tier contracts. What would be a large prime contract for a civil service center becomes a second-tier in an FFRDC. Surveyor, the lunar soft-lander, was one such contract. Hughes Aircraft held the contract with JPL. But when serious delays occurred and costs escalated, NASA didn't have direct control of this project even though it was of great consequence to the Apollo manned lunar landing. The situation deteriorated, and ultimately, NASA's senior management had to take strenuous action.

NASA Program Offices

NASA program offices were of necessity in Washington, DC. They had to be close to NASA's functional and administrative offices, as well as the executive office and Congress. Much of the congressional testimony was provided by program personnel. The nature of the program offices must reflect the contemporary responsibilities of the Agency. There shouldn't be too many program offices; their assignments shouldn't overlap any more than necessary; and project execution should, wherever possible, be conducted by their assigned Centers.

In 1967, there were three program offices, namely, Advanced Research and Technology, Manned Space Flight, and Space Science and Applications. The application projects included meteorology and communications, which were at one time separate from space science activities, but the projects were assigned to Goddard Space Flight Center, as were many of the space science projects. Priorities between science and applications were a potential source of conflict at the Center when it reported to two different program offices.

The communication and tracking projects were combined into a program office without line responsibility. The office had no control of any Center, but it did have direct responsibility for all surface networks. The antennas ranged from small to medium for Earth orbiting, as well as manned and unmanned vehicles, to large 220-foot antennas for manned lunar missions and distant unmanned planetary probes. Both Goddard and JPL had project assignments for these activities.

Keeping the Trains on Track

As discussed earlier, an attempt was made to obtain systems capability in the Manned Space Flight program office. For a variety of reasons, NASA was unsuccessful in recruiting sufficient numbers of individuals in this high-priced and scarce field. The Bell Laboratories of AT&T established a small subsidiary, Bellcom, to assist in this area, but there was an alternative long-term solution: independent, nonprofit corporations could be established for this purpose. Examples included the Aerospace Corporation to assist the Air Force's missile and space commands and the MITRE Corporation for support of Air Force ground capabilities. Other DOD and non-DOD agencies had similar arrangements. These Federal Contracted Research Centers (FCRCs) conducted much of the planning, engineering, and monitoring, but the contracting was handled directly by the matching government entity. In Jim Webb's mind, the use of a for-profit entity provided highly desirable flexibility, hence the use of Bellcom, a subsidiary of AT&T's Bell Laboratories. Although most of the technical personnel had to be hired, the internal administration was transferred in part from the Bell Laboratories, and there was always wise advice available at the Bell central headquarters. Finally, there was no NASA commitment or responsibility for the future. Of course, using a for-profit can lead to conflicts if the corporation is also bidding for hardware contracts in the same area where the company is providing systems advice. Conflicts of interest, or their appearance, can be ameliorated by exclusion clauses, but major corporations are reluctant to be so constrained. Since AT&T was not an aerospace corporation, no conflict developed, but caution was still required by both parties.

Day-to-Day Chores

During the six and a half years when I was general manager, I had a variety of chores that would have to be executed regardless of the Agency's organizational structure. The mail alone was a huge production. An executive in my office would determine who in the organization might need to see a piece of mail before I did. By the time a letter got to my desk, it might be in a big folder attached to memos from two dozen people in the organization, all of whom had ideas about how I should respond. Then, when my draft response had been written by me or by my assistant, there might be others I would want to review the response before it went out, so it might be recirculated with an endorsement attached, and each person would initial his approval.

I'm a great believer in face-to-face communication. That is one reason why I held project status reviews once a month for all NASA activities needing attention. It's also why I spent one quarter to one third of my time on the road, traveling to NASA Centers, contractor facilities, other government agencies, and foreign countries.

But the spoken word can be misunderstood, and sometimes all involved in an issue or project cannot be reached simultaneously. As a consequence, information systems and procedures are a necessity. Contained in this section are graphic samples of the systems and procedures used to manage NASA's programs during the time I was general manager.

Organizational Charts

The first clues about the interrelationships between individuals in an institution are provided by organizational charts, sometimes called wiring diagrams. A chart attempts to show who works for whom. From the standpoint of any individual, it provides an answer to the question, "Whom do I have to satisfy to remain on the job, to obtain a raise, or to be promoted?" Organizational charts also provide information in the large. How are projects, programs, and functions interrelated under the warm embrace of general management? Methods for management consumed a lot of Jim Webb's time and interest. In my first conversation with him, as I mentioned earlier, he asked me about my views on Montgomery Ward's mode of operation versus that of Sears Roebuck. The NASA that Jim inherited was nearly hierarchical, definitely in the Montgomery Ward mold (see figure 19). The Center Directors reported to the program directors, who in turn reported to the general manager (Associate Administrator) and, through him, to the Administrator. There were functional officers and staff, but authority flowed in a straight-downward chain of command from the general manager to the program offices to the project offices in the Centers. Incidentally, Doc Draper's Instrumentation Laboratory, where I spent my first professional years, was purely hierarchical even with individual projects having many of their own functional units.

After extensive discussion internally and with outside advisors such as Simon Ramo, Rube Mettler, and Art Malcarney, the organization was decentralized on 1 November 1961, more along the lines of Sears Roebuck (see figure 20). Note that the nine Center Directors and the five program directors reported to the Associate Administrator. Program resource allocations were authorized by the Associate Administrator with the support of the respective program director. The management activities of the Center Directors were also supervised by the Associate Administrator, with the assistance of the Director of the Administrative Office. This structure was introduced to provide more flexibility in the assignment of projects to the Field Centers and to provide the core management (the Triad) with greater visibility and control (see figure 2 in chapter 3). Note that the Center Directors became "two-hatted"; that is, in addition to satisfying the Triad that they were running a "taut ship," they also had to satisfy the program directors who were providing their project assignments and project resources.

I attended Keith Glennan's last management retreat, held in October 1960. Jim Webb held a similar retreat in Luray, Virginia, soon after he became Administrator. However, he felt that the setting could be subject to congressional criticism and that the retreat kept key managers out of touch with their organizations for too long. Henceforth, retreats were held at NASA centers. This had the added virtue of providing managers with an opportunity to become more familiar with other parts of the organization. At one such meeting, held at the Langley Research Center in 1963, the manned spacecraft team of Holmes, von Braun, Gilruth, and Debus strongly objected to the organizational structure implemented in November 1961. Mr. Webb was so displeased by this confrontation that retreats

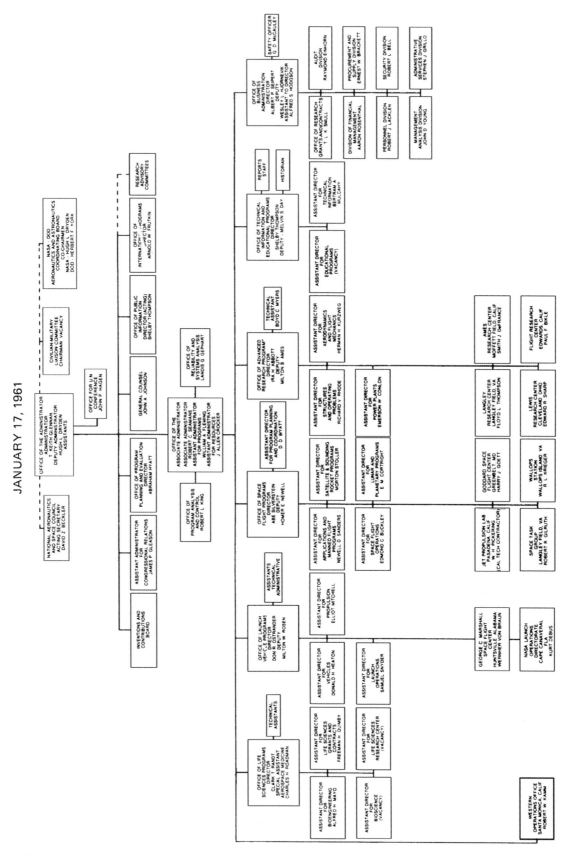

Figure 19. NASA organization during the last days of the Eisenhower administration, 17 January 1961. (Source: Jane van Nimmen and Leonard C. Bruno with Robert L. Rosholt, NASA Historical Data Book, Volume 1: NASA Resources 1958–1968 (Washington, DC: NASA SP-4012, 1988), p. 608.)

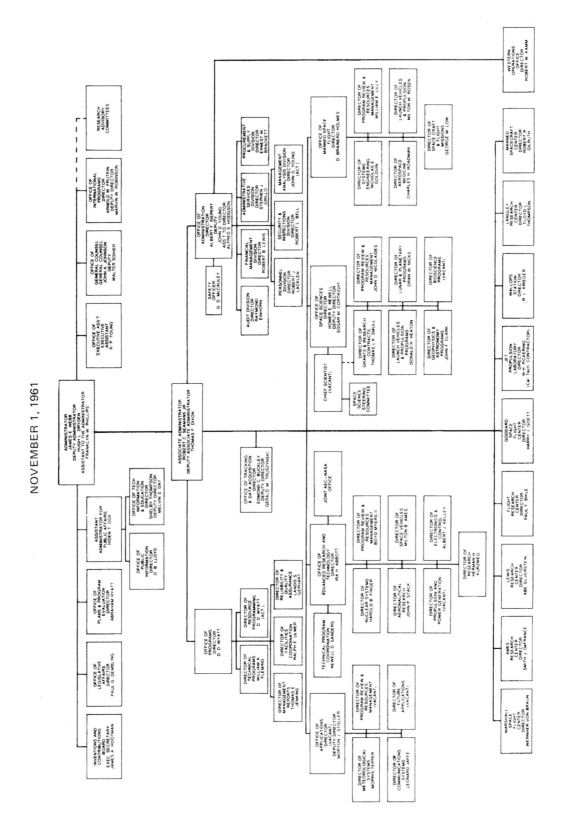

Figure 20. NASA organization as revised to conduct the manned lunar landing, 1 November 1961. (Source: Jane van Nimmen and Leonard C. Bruno with Robert L. Rosholt, NASA Historical Data Book, Volume 1: NASA Resources 1958–1968 *(Washington, DC: NASA SP-4012, 1988), p. 609.)*

were forever abandoned; however, soon thereafter, on the recommendations of many, including myself, there was a return to a more hierarchical organization (see figure 21). It should be noted, however, that the program offices and the Administrative Office continued to report to the Associate Administrator, thus maintaining their strong central role.

Soon after I became Deputy Administrator on 21 December 1965, NASA's functional offices (those reporting to the Administrator and those reporting to the Associate Administrator) were folded together, and all reported to the Office of the Administrator. This arrangement was discussed rather extensively at my Senate nomination hearing on 28 January 1966. The fourth organization chart (see figure 22) shows diagrammatically the changes that were made in and around the Office of the Administrator. Jim Webb was not happy with this arrangement and continued to prod me and Willis Shapley for improvements; meanwhile, he intro-

duced changes, such as a secretariat for the control of communications. I prepared a detailed, 18-page "eyes only" memorandum on the subject dated 16 December 1966. My recommendations apparently didn't satisfy Jim Webb's restless spirit, and they became less relevant to him in the wake of the Apollo 204 disaster.

In January 1966, when Harry Finger and his committees started reviewing the functional offices reporting to the Administrator, there were 12, ranging from General Counsel to Public, Defense, Legislative, and International Affairs. Many dealt with resources, including Personnel, Programming, Budgeting, Management Systems, Industry Affairs, University Affairs, Institutional Development, and Technology Utilization. Much discussion took place before a decision was reached to consolidate most of these resource functions into the Office of Organization and Management. Recently, in conversation with Harry Finger, who took over the function in

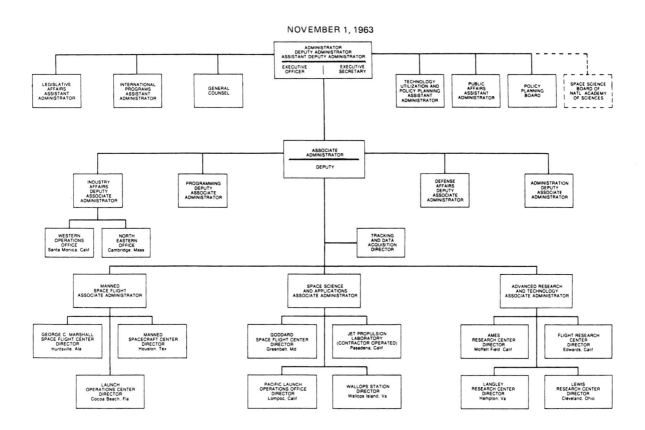

Figure 21. NASA organization as revised to strengthen Apollo and other project management teams, 1 November 1963. (Source: Jane van Nimmen and Leonard C. Bruno with Robert L. Rosholt, NASA Historical Data Book, Volume 1: NASA Resources 1958–1968 *(Washington, DC: NASA SP-4012, 1988), p. 610.)*

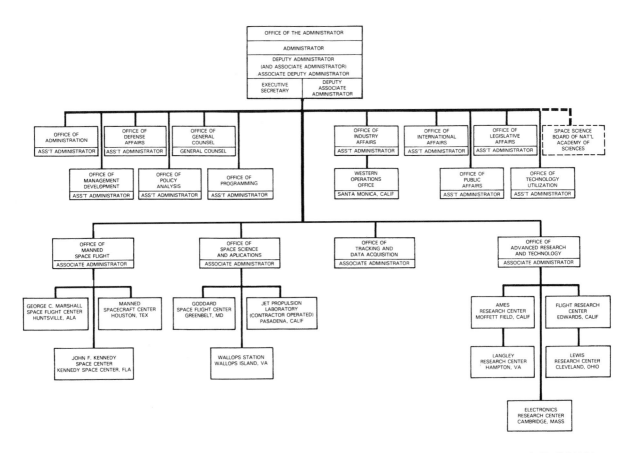

Figure 22. NASA organization following the consolidation of general management, 2 January 1966. (Source: Linda Ezell, NASA Historical Data Book, Volume 2: Programs and Projects, 1958–1968 (Washington, DC: NASA SP-4012, 1988), p. 617.)

March of 1967, he told me of his many meetings on the subject with Jim Webb. Jim finally asked him if he felt comfortable taking on this broad management role. He answered that he felt somewhat uneasy. When Jim asked why, Harry answered, "Because you're so involved in them." Jim said, "No, I want you to handle them."

Procurement

The preceding organization charts might imply that all decisions at NASA were made by the Administrator. Obviously, this would be an impossibility. Organization charts can only partially reveal the extent of decentralization within an organization.

In an organization, many functions must be performed—some in administration, some in programming and budgeting. Which officers are involved in establishing new top-level positions, in hiring personnel, in solving adjustments, in promotions, in disciplinary actions? How are audits conducted? How are safety and security inspected and

ensured? How are grants made to universities? "Who," "which," and "how" can be defined by writing detailed job descriptions, and much NASA time was expended in this process.

For myself, I've found that flow charts (figure 23) are much easier to comprehend. NASA projects were managed primarily at the Field Centers. The project manager at the Center prepared the procurement plan, an extensive document, which then passed through the Center Director and the program director, before a review by the procurement officer at Headquarters. Prior to 1966, final approval was given by the Associate Administrator. Afterward, plans came to the Office of the Administrator.

Once a plan was approved, the project manager prepared a request for proposal (RFP), which, if approved, was released by the Headquarters procurement office. The source evaluation board named in the procurement plan had both line and staff members, picked for their individual expertise, who participated in the evaluation. Their effort started when the RFP went out to the contractors.

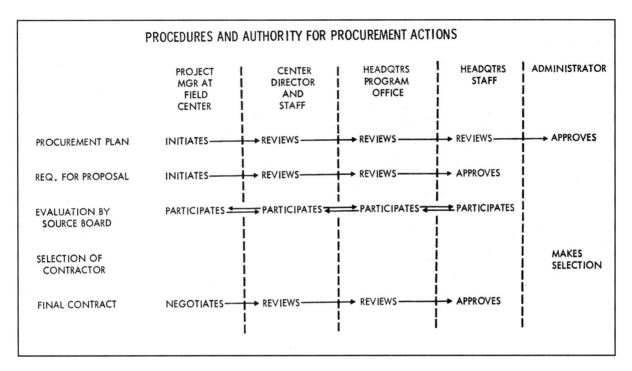

Prior to the submission of proposals, the board had to establish the basis to be used for scoring them. During the process, strict rules were enforced regarding communication with the contractors. If a contactor asked for clarification, all contractors were informed of the answer. If onsite inspection of one contractor was desired, all contractors were inspected. I fenced myself off from the evaluation so that I could remain immune from questioning by Congress or the media until the final day when the selection was made.

Once the team had prepared its findings, Webb, Dryden, and I would sit at the head of a table and the team would make its presentation. If the project was something coming out of Marshall, Wernher von Braun would be there watching, though he would have no say in the meeting. Our procurement people would be there as well. Webb used such meetings as a way of educating NASA, as well as a way of looking for hidden agendas. If anybody was trying to steer the project toward a particular contractor for whatever reason, we would try to smoke it out.

Afterward, the three of us would go into Webb's office with our chief procurement person, Ernest W. Brackett, and with Wernher (or his counterpart from the interested Center). "Okay," Webb would say, "we've heard the results of the source evaluation

board, now we'd like to hear from you, Wernher. What wasn't considered? Is there anything that was left out that you feel is important?" When he and Ernie Brackett had their say, they would leave, and the three of us were left with the decision.

As the junior person, I always went first, "Okay, Seamans, how do you look at this?" We would discuss it back, forth, and sideways, as Hugh and Jim advanced their views too. Finally Webb would say, "Okay, whom do you think we ought to pick, Bob?" I would tell him and why. Then Hugh would have his turn. If Hugh concurred with me, Jim usually agreed, and then the decision was made. Otherwise, there would be further discussion until an agreement was reached. In a few cases, more information was requested. The morning following a decision, Webb's executive secretary would have prepared a one-page decision paper saying that the Administrator of NASA had selected, say, North American Aviation for negotiation for the second stage of Saturn and giving reasons for the selection. All three of us would sign it. The press release would be based on this document, but the document itself was kept on file at NASA in case there was ever a congressional investigation.

When an agency like NASA procures buildings and other facilities, the contractors are provided with specifications and designs in the request for quotation (RFQ), and sealed fixed-price bids must be submitted by the contractors. The low bidder is selected if deemed competent, and a fixed-price contract is signed before work commences. The Army Corps of Engineers oversaw all of NASA's major construction on a sealed-bid basis. Most of the work for which NASA contracted directly was for research and development, where fixed-price contracts are not suitable. There must be pricing flexibility to accommodate unexpected factors. Hence, contractors were selected for negotiation. They were given a letter contract to commence work in anticipation of a definitive contract later. It was common practice to negotiate cost plus fixed fees. When there was a cost overrun, the government covered the contractor's extra expense, but the fee from which the contractor derived profit remained constant.

NASA started experimenting with incentive and award fees soon after Jim Webb became its Administrator. When the incentive is tied to cost, the contractor shares the cost of overruns with the government (the contractor's fee decreases). If cost savings occur, the contractor's fee is increased in proportion. Incentives can be related to schedules and performance as well as to cost. However, it's important not to allow the incentive arrangements to be overly complex, because then the government can lose control. The contract can so circumscribe the contractor's actions that the government cannot make needed contractual changes without it appearing to be a "golden handshake" (to let the contractor off the hook). Award fees are preferable in such circumstances. The award is made by the government against criteria agreed upon in advance. Such arrangements place highly desirable constraints on both the government and the contractor. A large percentage of NASA's Apollo business was conducted on either an incentive or an award-fees basis.

Budgeting

The budget is key to all government programs, and for that reason, the budget process is both important and time-consuming (see figure 24). There are periods when an agency must focus on three budgets. While using the budgetary resources provided in a given year, say fiscal year 1961 (FY61), the agency may be presenting the administration's request to Congress for FY62 and simultaneously negotiating the FY63 budget with the Office of Management and Budget and, ultimately, with the President. (The federal budget for any fiscal year currently runs from October 1 of the previous calendar year to September 30 of the year in question, but in the 1960s it ran from July 1 to June 30.)

On the day the new budget was released, the media attacked. All over Washington, correspondents nosed into the different agencies to find out what was important in each budget. We always had a press conference, which I conducted. As many as a hundred correspondents would be present. There might be TV reporters if the subject was sexy enough. I would run through what was novel in the budget, a process that might take 2 hours. Then there were questions.

I got a fair amount of scar tissue (figuratively speaking) from my years in government, and a fair amount of that came from the media. The intense media interest in the space program was a shock to me. I liked working with many members of the press. I understood that I could get gored, but I tried my best to have a good relationship with them. Most of them were pretty interesting people and fun to chat with, but I had to be very careful.

Bill Hines, the syndicated columnist who had called me "Moon czar," was particularly brutal to NASA. He would stand up and fire questions at us in a nasty, incisive way. Why were we so plodding? Why weren't we moving faster? Why weren't we more imaginative? When I came home Thursday nights, Gene would not let me read his syndicated articles until after dinner. Or, if they were too derogatory, she served me a martini first, which helped some.

I remember asking John Finney of the *New York Times*, "Why can't you do a positive, upbeat kind of story on NASA once in awhile?" His answer was, "Okay, I write a good article, and if I'm lucky it will be on page 33. If I write something controversial, I have a chance of getting it on page 1. It's as simple as that. I'm paid by what page I get my articles on."

History of NASA's 1962 Fiscal Year Budget

NASA's FY62 budget (see figure 24) was complicated by the changes that occurred starting with the Eisenhower submission, followed in March by a

NATIONAL AERONAUTICS AND SPACE ADMINISTRATION

Chronological History of the FY 1962 Budget Submission
(In thousands of dollars)

SUMMARY

ITEM	NASA Budget Submission	AUTHORIZATION — House Comm Action HR 6874 Rep No 391 5/12/61	House Comm Approved Budget	House Approved 5/24/61	Senate Comm Action HR 6874 Rep No 475 6/27/61	Senate Approved 6/28/61	Conf Comm Appd 7/19/61 Rep No 742 P.L. 87-98 7/21/61	APPROPRIATION — House Comm Approved Rep No 449 6/21/61	House Approved 6/7/61	Senate Comm Approved Rep No 620 7/25/61	Senate Approved 7/31/61	Conf Comm Appd 8/4/61 Rep No 850 P.L. 87-141 8/17/61
TOTAL APPROPRIATIONS:												
BASIC												
Research & Development	1,295,539		1,023,539	1,023,539	---	1,295,539	1,305,539	892,000[5]	892,000	1,278,000	1,278,000	1,220,000[10]
Construction of Facilities	262,075		139,075	139,075	---	262,075	252,075	116,250[6]	116,250	249,250	249,250	245,000
Salaries and Expenses	226,686		199,286	199,286	---	226,686	226,686	191,750[7]	191,750	221,750	221,750	206,750
SUPPLEMENTAL												
Research & Development	85,000	-85,000[1]			---			80,000[8]	80,000	85,000[8]	85,000	82,500[8]
Construction of Facilities	71,000[2]	---	71,000	71,000	---	71,000	71,000[2]	71,000[8]	---	71,000[8]	71,000	71,000[8][1]
GRAND TOTAL	1,940,300	3/	1,432,900	1,432,900	---	1,855,300	1,855,300	1,280,000	1,280,000	1,905,000	1,905,000	1,825,250
R&D Appropriation:												
BASIC												
Support of NASA Plant	89,110		77,110	77,110	---	89,110	89,110					
Research grants and contracts	7,600		7,600	7,600		7,600	7,600					
Life sciences	20,620		8,620	8,620		20,620	20,620					
Sounding rockets	9,000		9,000	9,000		9,000	9,000					
Scientific satellites	72,700		77,700	77,700		72,700	72,700					
Lunar and planetary exploration	159,899		103,899	103,899		159,899	159,899					
Meteorological satellites	50,200		40,200	40,200		50,200	50,200					
Communications satellites	94,600		44,600	44,600		94,600	94,600					
Mercury	74,245		74,245	74,245		74,245	74,245					
Apollo	160,000		72,100	72,100		160,000	160,000					
Launch vehicle technology	27,000		23,000	23,000		27,000	27,000					
Launch operations development	1,500		1,500	1,500		1,500	1,500					
Spacecraft technology	10,360		10,360	10,360		10,360	10,360					
Solid propulsion	3,100		18,100	18,100		3,100	10,200					
Liquid propulsion	93,020		78,020	78,020		93,020	93,020					
Electric propulsion	6,800		9,700	9,700		6,800	9,700					
Nuclear systems technology	36,000		36,500	36,500		36,000	36,000					
Space power technology	5,500		5,500	5,500		5,500	5,500					
Scout	3,675		3,675	3,675		3,675	3,675					
Delta	2,900		2,900	2,900		2,900	2,900					
Centaur	56,400		56,400	56,400		56,400	56,400					
Saturn	224,160		224,160	224,160		224,160	224,160					
Nova	48,500[4]	4/	4/	4/		48,500	48,500					
Tracking and data acquisition	38,650		38,650	38,650		38,650	38,650					
TOTAL R&D (BASIC)	1,295,539		1,023,539	1,023,539	---	1,295,539	1,305,539	892,000	832,000	1,278,000	1,278,000	1,220,000

Note (House Comm Approved Budget column, R&D section): House and House Comm acted on previously submitted requests which were different for some programs and projects.

Figure 24. Congressional budget history, NASA FY 1962. (Source: Robert C. Seamans, Jr., papers, MC 247, Institute Archives and Special Collections, MIT Libraries, Cambridge, MA.)

supplemental request to cover an expanded launch vehicle effort, and then a major budgetary expansion in May to initiate Kennedy's goal of manned lunar landing within the decade. NASA's total submission (budget request) is shown in the first column of figure 24. Notice that the submission covers three categories of activity. Two of these, "Research and Development" and "Construction of Facilities," are so-called "no-year" items; that is, the moneys can be expended over several years. Money in "Salaries and Expenses" is one-year money; it doesn't carry over to subsequent years.

The House and Senate receive a budget request in the form of voluminous reports. They commence committee hearings once their staffs have digested the material. The House, with over five times as many members, breaks its space committee into subcommittees in order to take a finer grained view of NASA plans and estimates. The markup of a bill takes place in a closed session and is then reported out to the appropriate congressional body, where open debate takes place on the floor of the House and Senate with the committee chairmen as protagonists and the committee members in a supporting role. I have had the unusual experience of sitting on the floor of the Senate assisting Senator Kerr in such a debate. I have also spent time outside the House Chamber and in the House Gallery, making myself available for questioning. Once the House and Senate bills have been approved, negotiation between the two bodies proceeds, culminating with closed sessions of the conference committee made up of the senior members of the two authorization committees.

But even after the President has signed an authorization bill, it doesn't provide dollars. First it must survive the appropriation process. Appropriation is similar to authorization, but its emphasis is cost. A new laboratory might be included in the cost of facilities, but if Congressman Albert Thomas, Chairman of the Independent Agencies Appropriation Subcommittee, were presiding, he might want to know why the cost covered special clean-air filters and other high-cost items. The committee might, as a result of these issues, reduce the estimated cost per square foot of the authorized building. Ultimately, an appropriation bill is negotiated and passed by both houses of Congress. However, even after a bill is signed by the President, the Office of Management and Budget holds the purse strings and may not release all funds to the agency.

There have been many instances when the appropriation bill has not been passed by the start of the fiscal year. In such cases, Congress passes a continuing resolution, which permits an agency like NASA to continue its existing programs, but not to initiate new projects. Finally, within limits prescribed in the authorization bill, the agency is given the opportunity to reprogram funds from one line item to another. The reprogramming has to be formally reported to Congress.

Scheduling

Anyone who has ever remodeled a home recognizes that the effort requires an interleaving of contractors. A delay by one has a domino effect on the others. Many separate schedules determine the final outcome.

Figure 25 is a Level 1 chart showing the actual launch dates of the six Gemini missions to date (filled-in arrows), as well as the scheduled launch dates of the remaining six missions (open arrows). Note that Gemini 6 and 7 were both planned for December 1965, within 10 days of one another. The "76" mission, as it came to be called, has already been discussed earlier in this book.

NASA also used Level 2 and Level 3 charts, which showed in greater, and still greater, detail the sequencing of the many work packages that had to be completed in order to achieve a launch. Such charts are essential for those with hands-on responsibility, but they do not reveal whether a project is staying on track or whether, as happens most often, deadlines are "slipping" (constantly being moved toward a later date). Such slippages usually result in escalating costs.

To stay alert to and (hopefully) prevent slippages, I developed and implemented "trend charts" (see figure 26) like the one drawn for a hypothetical mission. If, on 1 January 1961, a launch had been scheduled for November 1963, and if that launch had stayed on schedule throughout the reporting periods in 1962 and 1963, progress would be shown as a straight, horizontal line labeled "IDEAL." If, instead, the project kept slipping its deadlines such that the launch did not occur until some time in 1965, its trend would have followed a jagged line like the one labeled "ACTUAL." Note

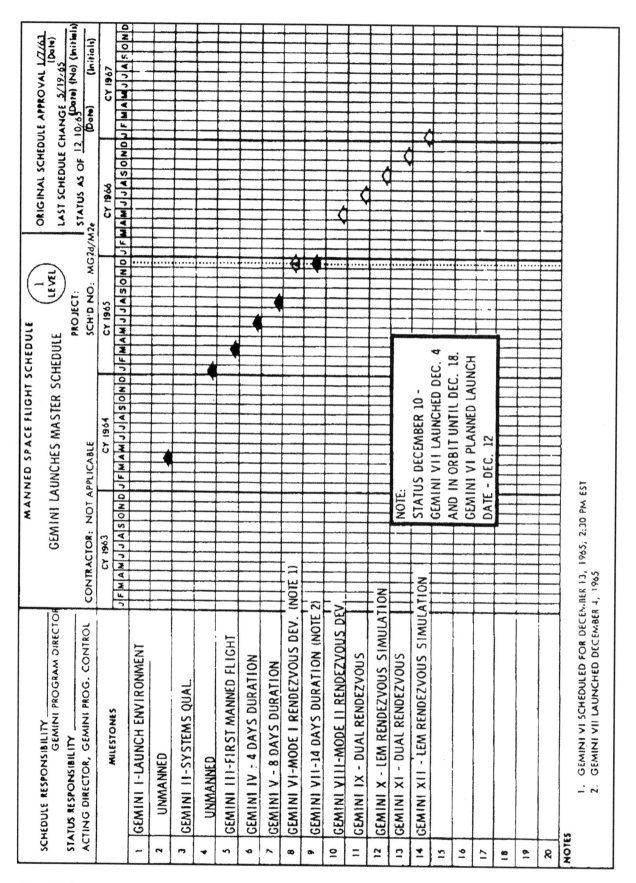

Figure 25. *Gemini master launch schedule on 10 December 1965, with Gemini 7 in orbit and Gemini 6 about to be launched. Five additional Gemini missions remained. (Source: Robert C. Seamans, Jr., papers, MC 247, Institute Archives and Special Collections, MIT Libraries, Cambridge, MA.)*

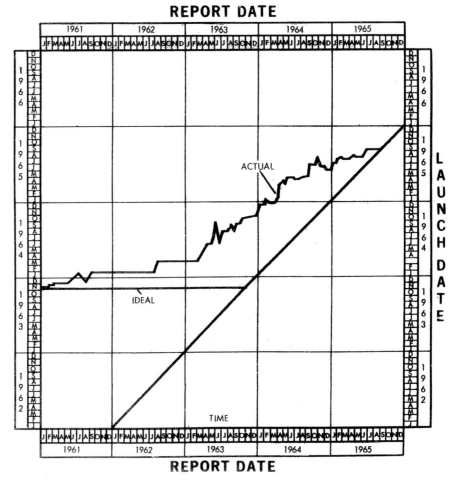

SCHEDULE HISTORY

LAUNCH DATE TREND CHART

LAUNCH DATE AS
PERIODICALLY ESTIMATED

Figure 26. This chart shows a hypothetical mission experiencing major delay. This type of chart was used to focus management on unfavorable project trends. (Source: Robert C. Seamans, Jr., papers, MC 247, Institute Archives and Special Collections, MIT Libraries, Cambridge, MA.)

that in this hypothetical case, the schedule held pretty well until March 1963, when a nearly day-for-day slippage began to occur, ultimately delaying the launch nearly two years.

The trend chart for the complete Gemini Program of 12 launches dated 31 October 1966 is shown in figure 27. Originally, the program was to be completed by June of 1963. However, after negotiations with McDonnell Douglas were complete and the contract was definitized, the final launch was scheduled for early 1967. It is a tribute to George Mueller and the entire Gemini government-industry team that the 12 launches were completed ahead of schedule.

Project Approval

When DeMarquis "D." Wyatt became the director of the Office of Programs at NASA, we agreed to use as simple a system as possible to release research, development, and facility funding.

We attempted to keep each Project Approval Document (PAD) less than a page in length and to use the same format for each document, one of my pet idiosyncrasies. The PAD included the project number, program, project name, and purpose. It also included the "level of effort" (budgetary limits) in thousands of dollars. Finally, a project might have a variety of special stipulations. An example is shown in figure 28.

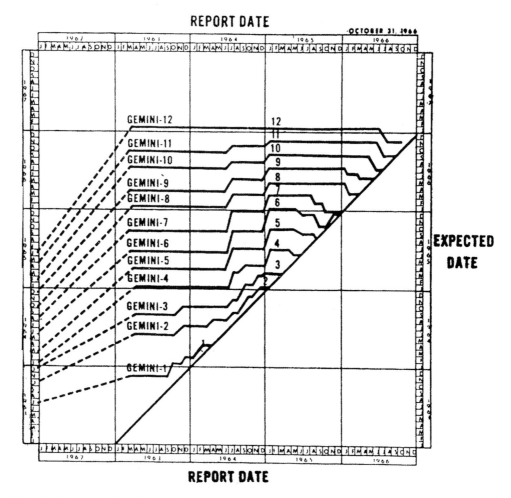

Figure 27. Trend chart for Gemini's 12 launches as of 31 October 1966, 11 days before the completion of the program. (Source: Robert C. Seamans, Jr., papers, MC 247, Institute Archives and Special Collections, MIT Libraries, Cambridge, MA.)

I don't know how many hundreds of PADs D. Wyatt and I signed, but the 72 PADs related to Apollo are contained in figure 29. PADs were the basis for the monthly status reviews that I held with each of the program directors. These reviews included an updating of costs, schedules, and performance, with emphasis on areas where deficiencies existed. The approval documents and the agenda for their monthly review resulted from interaction between D. Wyatt's office and the control group in each program office.

There were two regularly scheduled Headquarters sessions each month. I chaired the Project Status Reviews and Jim Webb the Program Review. I'm a strong believer in management oversight and correction on a monthly basis. On a quarterly time scale, details can be blurred between meetings, and it

may become too late to head off impending disaster. I also believe that key line personnel must be present, as well as representation from the functional offices. When the meeting was on manned space-flight issues, either George Mueller or the Gemini and Apollo directors, Chuck Mathews and Sam Phillips, were present. Prior to the meeting, I had a session with D. Wyatt and his program office to review the agenda and the critical areas that needed discussion. Soon after the meeting, the action items would be documented, signed by me, and sent to the appropriate managers.

The purpose of the Program Reviews was to bring Mr. Webb and NASA's leadership up to date on NASA's entire effort. The sessions consumed an entire Saturday and were repeated on Mondays for other government agencies. These meetings also

December 18, 1961

Project Number: 32-0304

Program: Apollo

Project: Apollo Spacecraft

Purpose: Development of a three man spacecraft capable of
                   achieving manned lunar landing through direct ascent or
                   orbital rendezvous techniques utilizing a modular concept.

Estimated Level of Effort:

    a. FY 1961 $ --
    b. FY 1962 Budget: $52,000
    c. FY 1962 Approval: Command and Service Module $42,500
                             Instrumentation and Scientific Eq. $ 2,100
                             Operational Support $ 1,500
                             Supporting Development $ 1,675
                             Guidance and Navigation $ 4,500
                                                                 ─────────
                                      Total $52,275 *

Project Stipulations:

    a. Project Management: Manned Spacecraft Center
    b. Program Management: Office of Manned Space Flight
    c. Scope and content of Project Apollo shall be consistent with PPDP
        with amendments of October 23, and November 1, 1961, submitted by
        the Office of Manned Space Flight less the exclusions noted under
        (e).
    d. Utilization of NASA-PERT and the NASA Financial Management
        Reporting System for Cost Type contracts will be utilized. PERT
        events and Financial Management Reporting System cost reporting
        categories shall be selected in a manner which will permit
        integrated time/cost management control and reporting.
   *e. Funds are not approved for the high-energy abort and lunar take-off
        propulsion development included under Supporting Development ($925)
        pending recommendation on management of this work.
   *f. Funds are not approved for lunar landing propulsion development.

Project as specified above approved

CONCUR _____

APPROVED _R. C. Seamans, Jr._____

*Indicates change

Figure 28. Project Approval Document (PAD) for the Apollo spacecraft, 18 December 1961. (Source: Robert C. Seamans, Jr., papers, MC 247, Institute Archives and Special Collections, MIT Libraries, Cambridge, MA.)

APOLLO PAD CHRONOLOGY
(Items marked with * are summarized on Chart)

Date	Title	Date	Title
*7-20-61	Saturn C-1 Development	6-7-62	Advanced Saturn
*7-20-61	J-2 Engine Development	6-19-62	Advanced Saturn
*7-20-61	F-1 Engine Development	7-27-62	Apollo Spacecraft
8-1-61	Apollo-Adv. Tech.	*9-4-62	Saturn C-IB
*8-9-61	Apollo Spacecraft G&N	9-21-62	F-1 Engine
*9-26-61	Saturn C-3	9-21-62	J-2 Engine
10-18-61	Saturn C-1	9-21-62	Vehicle Procurement
10-25-61	J-2 Engine	9-21-62	Saturn C-5
*11-3-61	Apollo Spacecraft	9-21-62	Saturn C-1
11-3-61	Apollo-Adv. Tech.	9-21-62	Apollo Spacecraft
11-7-61	Saturn C-1	11-13-62	Saturn C-5
11-16-61	Apollo-Manned	11-13-62	Vehicle Procurement
	18 orbit mission	11-28-62	Vehicle Procurement
*12-18-61	J-2 Engine	12-11-62	Vehicle Procurement
*12-18-61	Apollo Spacecraft	1-14-63	Apollo Spacecraft
12-22-61	Advanced Saturn	2-28-63	Apollo Spacecraft
*1-2-62	J-2 Engine	3-12-63	Vehicle Procurement
1-2-62	Advanced Saturn	4-4-63	Integration & Checkout
*1-25-62	Advanced Saturn	4-4-63	Aerospace Medicine
1-26-62	Apollo Spacecraft	4-4-63	Systems Engineering
*1-29-62	Apollo Spacecraft	*6-30-63	Apollo Spacecraft
3-13-62	Advanced Saturn	*6-30-63	Vehicle Procurement
3-13-62	Apollo Spacecraft	*6-30-63	RL 10 Engine
*3-21-62	Saturn C-1	*6-30-63	H-1 Engine
3-29-62	Apollo Spacecraft	*6-30-63	F-1 Engine
3-30-62	Saturn C-1	*6-30-63	Launch Operations Suppor
4-9-62	Advanced Saturn	*6-30-63	Launch Instrumentation
4-25-62	F-1 Engine	*6-30-63	Systems Engineering
4-25-62	J-2 Engine	7-10-63	J-2 Engine
4-26-62	Advanced Saturn	9-10-63	L/V Supp. Tech.
4-26-62	Saturn C-1	9-10-63	Propulsion Supp. Tech.
*4-27-62	Apollo Spacecraft	9-10-63	Launch Ops. Supp. Tech.
*4-27-62	Vehicle Procurement	*6-3-64	Apollo Total
5-2-62	Advanced Saturn	*8-28-64	Apollo Total
5-21-62	Saturn C-1	*12-16-64	Apollo Total
6-4-62	Apollo Spacecraft	*8-18-65	Apollo Total
6-7-62	Saturn C-1		

Figure 29. Project Approval Documents for the Apollo Program. (Source: Robert C. Seamans, Jr., papers, MC 247, Institute Archives and Special Collections, MIT Libraries, Cambridge, MA.)

provided Jim with an opportunity to educate NASA on his expectations, and he was voluble and not bashful. Sometimes I'd receive a call with the words, "I don't know what he's talking about." I hope my response was helpful, but I would always add, "In the future, you better listen carefully."

Apollo Management

This section includes a discussion of the organizational evolution during the sixties. When George Mueller became the Associate Administrator of Manned Space Flight, the Apollo and Saturn project offices soon became two-hatted, as shown in figure 30.

Lieutenant General Samuel Phillips was assigned to NASA to serve as the director of the Apollo Program Office at the request of George Mueller, Associate Administrator for Manned

Space Flight. The project directors (Apollo, Saturn I/IB, Saturn V, and Launch Operations) all reported to Sam Phillips on Apollo matters. Each had five functional offices for program control, systems engineering, test, reliability-quality, and flight operations. These functional offices communicated directly with each other, but decisions at each Center were made by the project directors. Meanwhile, it was the responsibility of the Center Directors (in Texas, Alabama, and Florida) to provide an appropriate institutional environment for the projects. Their duties included the allocation of personnel. Also, the Center Directors had Apollo Program input through George Mueller. When Sam Phillips reported to George periodically, the Center Directors were present, serving as members of his Apollo Board of Directors.

I never attended George's Apollo reviews, just as Jim Webb never sat in on my Project Status Reviews. However, their timing and brutal detail

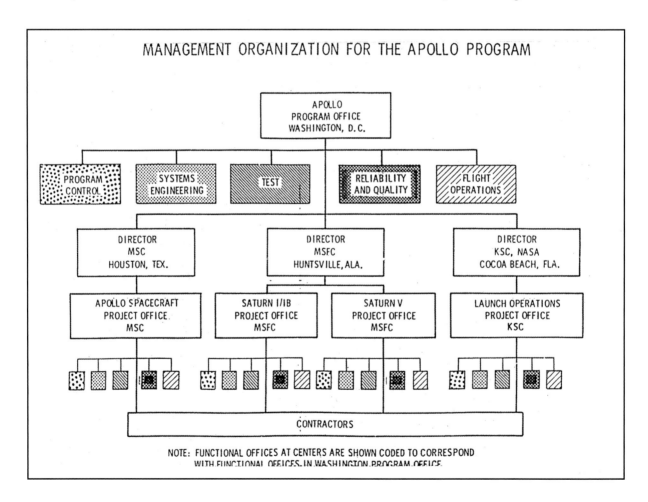

Figure 30. Management organization for the Apollo Program. (Source: Robert C. Seamans, Jr., papers, MC 247, Institute Archives and Special Collections, MIT Libraries, Cambridge, MA.)

were well known. All of his key players were hard-working, with extensive responsibilities, but George was indomitable. He didn't believe in weekends. He often called for meetings on Sundays, and many times the material presented was nicknamed "pasteurized." That is, the information was so extensive that late on a weekend, the participants' ability to absorb was waning, and the charts were merely "past your eyes." But not to George. Even at day's end, he was direct and incisive.

It was a privilege for me to know the individual members of George's council. They attended both Headquarters Project and Program Reviews, but in addition, I was often with them while traveling, exercising, or visiting in their homes. George lived next door and made excellent dry martinis. Joe Shea was an excellent tennis player, and we climbed many fences into locked tennis courts to play before sunrise. I nearly broke my hip skateboarding on George's cement driveway. Bob Gilruth and I loved

boats. He built his in a shed next to his house. Wernher loved the outdoors and his houseboat on the Tennessee River. His daughter went to the Cathedral School and joined us at mealtime on occasion. I often shared the podium with Kurt Debus at the Cape. Sam Phillips was quiet, conscientious, and persevering. I was fortunate to work with him, first at NASA and later when we both held critical responsibilities in the Air Force.

The five functions described previously—program control, systems engineering, testing, reliability and quality, and flight operations—permitted NASA to have centralized management at Headquarters for overall control of the Apollo Program. Sam Phillips was its arbiter, decision-maker, and spokesman. However, the information he received and the actions he disseminated were distributed among Headquarters and the Field Centers. In this way, key decision-makers at Headquarters availed themselves of the technical

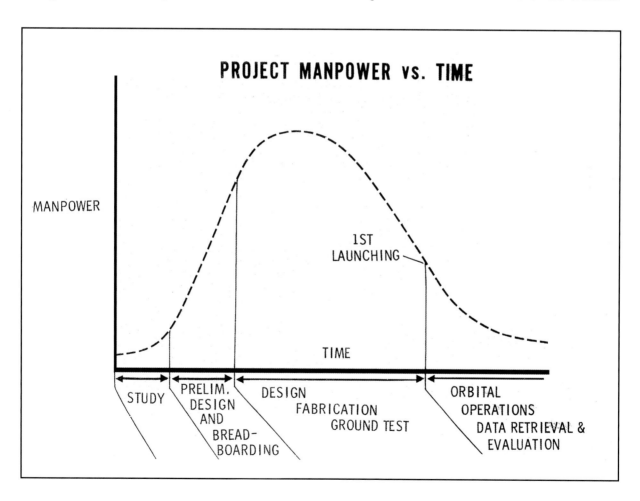

Figure 31. Manpower requirements during the advancing phases of a program. (Source: Robert C. Seamans, Jr., papers, MC 247, Institute Archives and Special Collections, MIT Libraries, Cambridge, MA.)

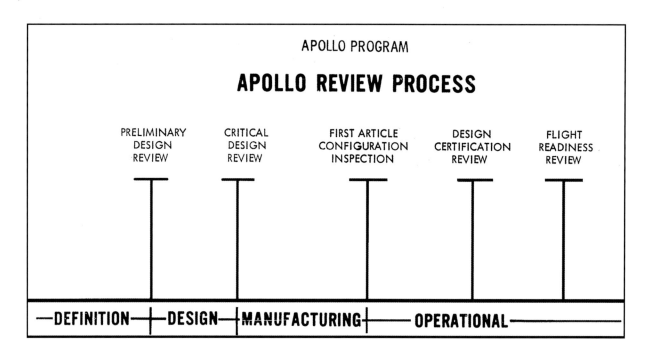

APOLLO PROGRAM

APOLLO REVIEW PROCESS

PRELIMINARY DESIGN REVIEW

CRITICAL DESIGN REVIEW

FIRST ARTICLE CONFIGURATION INSPECTION

DESIGN CERTIFICATION REVIEW

FLIGHT READINESS REVIEW

—DEFINITION—|—DESIGN—|MANUFACTURING|——OPERATIONAL——

Figure 32. Apollo Review Procedures, the essential milestones. (Source: Robert C. Seamans, Jr., papers, MC 247, Institute Archives and Special Collections, MIT Libraries, Cambridge, MA.)

competence and knowledge at the Centers, and the project directors in the Centers were kept current on Headquarters activities. Aaron Cohen tells of an early assignment at the Johnson Space Center. Joe Shea, the Apollo project director, put him in charge of the Interface Control Documents (ICDs). Joe then escorted Aaron to Marshall Space Flight Center to meet Wernher von Braun. Joe advised Wernher of Aaron's responsibilities and Wernher asked, "What are ICDs?" When informed, Wernher bought in, and Aaron estimates that he eventually negotiated over 1,000 such documents.

The management of Apollo was clearly disciplined and distributed. There was no large central authority issuing detailed instructions. Rather, there were interface documents that were continually updated to correspond with developments at the Centers. Final development cannot take place without the interfaces, and interfaces require knowledge of the developed hardware. Progress can only take place in an iterative fashion with strong Center participation.

It's also important to keep in mind the progress of Apollo in the large picture. Figure 31 is a stylized representation of the phasing experienced in such large projects. Prior to President Kennedy's ringing endorsement, there was a series of discussions and studies internal and external to NASA. Once NASA

received its mandate for the manned lunar landing, it expanded its workforce, as did other government agencies. But the largest increase in manpower was contracted to industry. Apollo started with preliminary designs and the breadboarding of hardware. By the middle of the decade, designs were well along and the fabrication of hardware was rapidly advancing. Prototype hardware was needed for thorough testing of performance in high-stress environments. This ringing out of the hardware was coupled with the extensive planning that was required for operational testing, first on the ground and ultimately in space. Of course, Apollo had several flight missions prior to the lunar landing. The first vehicle test was followed by manned flights in Earth orbit and around the Moon before the Apollo Program achieved the ultimate goal of lunar landing. By the time the goal was reached, the total manpower on Apollo was greatly reduced, but funds were still required for wrapping up the program.

Each major Apollo project passed through a review and approval process (see figures 31 and 32), making its way progressively through definition, design, manufacture, and flight operations. The key reviews were design certification and flight-readiness. During the former, results of all major tests were described and discussed. Such tests included static-engine and stage firing, vacuum testing on spacecraft, and electrical tests of checkout equipment. The flight-readiness reviews were all-encompassing and included presentation of the results of operational tests of the ground, ship, and airborne network, along with Launch Control, Mission Control, and the flight hardware. If a few items were outstanding, they would have to be cleared before the time of the launch. Sam Phillips was the chairman of the committees overseeing all key project reviews and approvals.

Saturn I was developed in the then-normal fashion, with four flight tests of the first stage before testing of the second stage. When George Mueller became the Associate Administrator for Manned Space Flight, he introduced NASA to "all-up" system testing. When the Apollo/Saturn V was launched for the first time in December 1967, all three launch vehicle stages were operational, as were the spacecraft and service module. All worked satisfactorily. Such success could not have been achieved without the extensive testing that culminated in a completed flight-readiness review.

Soon after the first Saturn V launch, I retired from my official capacity, and I was sworn in as a NASA consultant on 5 January 1968, the day following my retirement. From then on, I used the consultant's offices and only participated in management decisions on the few occasions when I was asked. I did review with Sam Phillips the design changes being made to the Apollo capsule to improve fire protection. I also spent time with Gene Emme, the NASA historian, on my exit interview. Finally, I participated in the UN conference on "Peaceful Uses of Space." Jim Webb was also there and told me of the plan to circumnavigate the Moon on the first manned Saturn V mission.

Chapter 6:
THE GRAND FINALE

At the start of 1968, NASA was rapidly converging on its goal of a manned lunar landing within the decade. Actually, success was only 19 months away—from January 1968 to July 1969. Prior to the landing, there would be two unmanned spaceflights and four manned flights.

Monday, 22 January 1968, Apollo 5, Unmanned

Apollo 5 was launched on 22 January 1968 with the primary objective of testing the lunar descent propulsion; the ascent propulsion, including restarting capability; the spacecraft structure; the instrumentation and control; and the second stage of the Saturn IB. After separation from the Saturn booster, the Lunar Module was in an elliptical orbit and proceeded with a test of the descent stage. The planned 39-second burn only lasted 4 seconds due to a computer program glitch. The ground controller shifted to an alternate plan and tested the descent stage first with a 26-second burn at 10-percent thrust and, finally, a 7-second blast at maximum thrust. Later, each stage was put through its paces, ending with an ascent-stage firing of over 6 minutes.

The descent engine was throttleable, like that of an automobile—the only rocket motor with this capability, which was necessary for a soft-landing on the lunar surface. Ignition of the ascent engine was essential to recovery; there was no redundancy. For this reason, hypergolic fuel was utilized—ignition occurred without the need for a separate firing source. In an abort, this stage had to ignite and fire even while the descent stage was still providing thrust. During the test, the so-called "fire in the hole" was successful. The flight was judged satisfactory, and the Lunar Lander was declared ready for manned flight.[1]

Thursday, 4 April 1968, Apollo 6, Unmanned

The second and final unmanned launch of Apollo occurred on 4 April 1968. The Apollo 6 was the second launch of the Saturn V. The first stage functioned as planned, but two of the second-stage J-2 engines shut down prematurely. To compensate, the remaining three engines automatically burned longer, as did the single J-2 engine in the third stage. The compensation was nearly perfect; however, it left the Apollo in a somewhat elliptical rather than circular orbit.

The third stage failed to reignite, so the Apollo capsule and service modules separated from the staging; then, by firing the service module, an altitude of nearly 14,000 miles was achieved at apogee. From there, a reentry speed of 22,400 mph was achieved, rather than the planned 25,000 mph. Although only four of the five flight objectives were achieved, the flight demonstrated a remarkable redundancy when two of the five J-2 engines flamed out, both on the same side of the rocket. If the engines had been on opposite sides, there wouldn't have been the imbalance that tended to topple the structure. When all the data were analyzed, the tests, coupled with the 100-percent success of the previous flights, were judged complete and the Apollo and Saturn V were judged ready for manned flight.[2]

In mid-August of 1968, I attended a U.N. conference on "Exploration and Peaceful Uses of Outer Space." The conference took place in Vienna, amidst the best coffee houses and chocolate in the world. I presented a paper at a panel on management issues related to manned space exploration. I was also chairman of the full-fledged assembly where each of the 74 nations participating could have five representatives present. The night before this session, the Soviet Army entered Czechoslovakia. The staff secretary for the morning meeting was a Soviet colonel, and a Czech professor was presenting a paper. I received a message from U Thant, Secretary of the U.N., to close the session if any political shenanigans occurred. Specifically, I was to say "meeting adjourned" and slam down the gavel.

Jim Webb was also at the meeting and was concerned about the recently announced "Intersputnik" that would be established by the Soviets and other Socialist countries to compete with Intelsat, the western satellite communication network with over 100 member nations. However, he invited me to his room at the Intercontinental Hotel to tell me of his recent telephone calls from Tom Paine, NASA's Deputy Administrator at that time. Tom proposed a circumlunar flight for the next Apollo mission. He advised Jim that there were indications of an early Soviet manned mission to the Moon, and the Lunar Module was not ready for a 1968 mission as previously planned. However, all the necessary elements were ready for manned circumlunar flight. When asked for my views, I first thought of the caveat that I wasn't up to date on NASA readiness. Then I said that at first blush, such a mission in 1968 seemed at the edge of the envelope. But I ended my comments with a reminder of my thoughts on EVA prior to the Gemini 4 mission. NASA should go when ready and also should attempt to accomplish as much as possible on each mission.

Friday, 11 October 1968; Apollo 7; Wally Schirra, Don Eisele, and Walter Cunningham

The first manned flight of Apollo occurred in October 1968 between the 11th and the 22nd. Wally Schirra was the commander, with Don Eisele and Walter Cunningham completing the three-man crew. The Apollo 7 spacecraft weighed 36,500 pounds and was carefully redesigned for safety. The two-piece hatch was replaced with a single one that was quick-opening. Also, there were extensive material substitutions to reduce flammability. All flight objectives were achieved. The service module engine was fired eight times, including the de-orbit burn. Although the crew was kept busy with time-consuming maintenance in addition to their regular duties, there was still time for photographs of Hurricane Gladys over the Gulf of Mexico and a long plume of air pollution over the United States.

1. *Astronautics and Aeronautics, 1968: Chronology of Science, Technology, and Policy* (Washington, DC: NASA SP-4010, 1969), p. 13.
2. Ibid., p. 77.

Figure 33. This view of the rising Earth greeted the Apollo 8 astronauts as they came from behind the Moon after the lunar orbit insertion burn. (NASA Image Number 68-HC-870, also available at http://grin.hq.nasa.gov/ABSTRACTS/GPN-2001-000009.html)

There were also five live TV broadcasts with both outside photography and in-capsule gymnastics and commentary. In one, the astronauts displayed a sign bearing greetings from "the lovely Apollo room high above everything." The astronauts won honorary membership in the American Federation of Television and Radio Artists. The astronauts were recovered by the USS *Essex* after circling Earth 163 times.[3]

Saturday, 26 October 1968, Soyuz 3

The Soviets launched Soyuz 3 from Baikonur Cosmodrome with a "powerful rocket booster" on 26 October, four days after the landing of Apollo 7. The Soviets conducted a variety of scientific, technical, and biological experiments; transmitted TV pictures; and conducted a rendezvous with Soyuz 2. Clearly, the Soviets had recovered from Komarov's

fatal accident and were proceeding aggressively with their manned space effort.[4]

Saturday, 21 December 1968; Apollo 8; Frank Borman, Jim Lovell, and Bill Anders

Sometime in early December, I received an invitation to fly in a visitors' plane from Washington, DC, to the Cape, for the launch of Apollo 8, with Frank Borman, Jim Lovell, and Bill Anders aboard.

On 19 December, two days before Apollo 8's liftoff, I received a call from Mel Laird. President-elect Nixon had introduced his cabinet nominees, including Mel Laird as Secretary of Defense, at the Pierre Hotel in New York. From that TV program, I recognized his name, but I was surprised when he asked me for lunch the following day at the

3. *Astronautics and Aeronautics, 1968: Chronology of Science, Technology, and Policy* (Washington, DC: NASA SP-4010, 1969), p. 250.

4. Ibid., p. 264.

Carleton Hotel in Washington. Could I join him at noon? The answer was yes, provided that I could get to National Airport by 4:00 p.m. I was met on arrival by Bill Baroody, Mel's assistant, who introduced me to a gigantic organizational chart of the Department of Defense. Mel arrived quite late, explaining that he had been meeting at the Pentagon with Clark Clifford, who would soon be giving Mel the key to the establishment. After much discussion and a good lunch, Mel asked me to join his team as Secretary of the Air Force. I was aghast. It wasn't possible. Gene was in the hospital; we had just bought a new house in Cambridge, etc. However, I agreed not to say no until the following week. I headed for the airport. To my surprise and subsequent pleasure, I was seated next to Jack Benny. We had a great conversation, which centered on the space program, as well as his concern for the health of Bob Hope. Actually, Bob Hope outlasted Jack Benny by several decades.

The next morning, 21 December, Apollo 8 lifted off from Cape Canaveral. The viewing stand was over a mile from the pad. Hence, the sound from Apollo took time to reach those sitting there holding their breath. First, however, a voice blared, "Ignition," followed, seconds later, by "Liftoff." Just as the sound of the five engines was surrounding us, Apollo was clearing its umbilical tower, as in previous Saturn V launchings. It's hard to describe the sound. It was overwhelming—it wasn't just heard; it was felt overall, from the low-frequency rumbles to the high-pitched crackling. This monster passed safely through maximum g's, where the dynamic air pressure was greatest, to first-stage shutdown and second-stage ignition, amidst much cheering. And then we watched intently until the Apollo 8 spacecraft disappeared from our view.

Nearly 3 hours after liftoff, the third-stage engine was fired, sending Apollo 8 on its translunar passage. Earth's gravity would slow Apollo on its lunar trajectory until lunar gravity exceeded Earth's pull; then, Apollo would accelerate as the Moon appeared to increase in size and resolution of detail. Only small corrections in speed were required en route: first, an increased speed of 24 feet per second (fps); then, a reduction of 2 fps to make the Moon's closest approach an altitude of 60 miles.

On Christmas Eve, the service module was burned for 4 minutes and 2 seconds, giving Apollo an apolune of 194 miles and perilune of 69 miles. The orbit was later circularized at 70 miles. Back home, with Gene hospitalized, our children were preparing for Santa's arrival both at home and near Gene's hospital bed. There was time, however, to listen to the crew read from the first chapter of Genesis and then wish all of us "Good night, good luck, a merry Christmas, and God bless all of you—all of you on the good Earth."

There were five TV transmissions from Apollo 8: some of Earth at 139,000 miles, some from 201,000 miles while en route to the Moon, and then some of the lunar surface "like dirty beach sand" of prospective landing sites, as well as of mountainous areas. But the most spectacular image appeared as Apollo 8 came from behind the Moon and the astronauts saw the blue Earth appearing to rise above the lunar horizon. Each astronaut grabbed for a Hasablat camera and one of them took the photograph (see figure 33) we have all seen many times. The beautiful but small spaceship, Earth, is there in its entirety, in sharp contrast to the desolate, dead lunar surface. This photograph is a graphic reminder, for all to admire, of the treasure we inhabit. If Apollo 8 inspires us to conserve our planet, it is worth many times the cost of man's lunar travels.[5]

Tuesday, 14 January 1969, Soyuz 4/Wednesday, 15 January 1969, Soyuz 5

In mid-January, we had another sharp reminder that the Soviets were not just treading water. Soyuz 4 was launched on 14 January 1969 and Soyuz 5 on 15 January. After the cosmonauts conducted a variety of experiments, they performed a rendezvous and then manually docked, but with no hatch between them. Two cosmonauts in Soyuz 5 put on their "special" spacesuits with a new "regenerative life-support system," left their spacecraft, and joined the cosmonaut in Soyuz 4. The Soviets demonstrated, in their words, "the world's first experimental cosmic station." Regardless of the hype, they had demonstrated the first transfer

---

5. Ibid., p. 318.

between spacecraft; they clearly weren't backing away from space exploration and, most probably, a manned lunar mission.[6]

Monday, 3 March 1969; Apollo 9; Jim McDivitt, Dave Scott, and Rusty Schweickart

On 3 March 1969, Jim McDivitt, Dave Scott, and Rusty Schweickart went through a series of complex maneuvers with Apollo 9. It was the first test of the Lunar Module with astronauts in space. The astronauts first flew in the Command Module for several days, performing housekeeping and separating from the third stage to redock cheek to jowl with the Lunar Module. On the third day, Jim and Rusty entered the Lunar Module using the on-board hatches. They conducted the first test of the lunar descent stage and returned to the Command Module. Rusty then spent 37 minutes outside the capsule using selected hand- and footholds to reach the Lunar Module and return.

On the fifth day, Rusty and Jim reentered the Lunar Module. They separated from the Command Module, which backed 3 miles away from the Lunar Module using the control-system thrusters. The Lunar Module then went through a simulated Moon landing, using the ascent rocket to separate from the Command Module by over 100 miles. After 6.5 hours of "time on the Moon," the ascent stage fired its engines, separating from the descent stage, and returned the astronauts to the Command Module for docking and return to Earth. Apollo 9 landed within sight of the USS *Guadalcanal* for pickup by helicopter. The first test of a manned Lunar Module was most successful, achieving all major objectives. NASA was now ready for lunar landing operations. President Nixon congratulated the crew and said that the mission showed "what man can do when they bring to any task the best of man's mind and heart."[7]

While Apollo 9 was under way, I received a telephone call from Mel Laird. By then, I was the Secretary of the Air Force. He told me the President had two candidates for the next Administrator of NASA, and one of them, a Democrat, was Tom Paine. He asked for my recommendation. I quickly answered that if the President wished to ensure a safe landing on the Moon within the decade, he'd nominate Tom as the Administrator. Tom's nomination was announced the following day (5 March). Jim Webb's strategy for ensuring NASA's continuity had prevailed. Jim was so political that he couldn't have survived in the Nixon administration for many days. By resigning in the Johnson administration, he cleared the way for Tom Paine.

Sunday, 18 May 1969; Apollo 10; Tom Stafford, John Young, and Gene Cernan

The Apollo 10 spacecraft lifted off Pad B of Complex 29 on 18 May 1969. Tom Stafford was the commander, John Young the Command Module pilot, and Gene Cernan the Lunar Module pilot. The flight went by the book, with minimal corrections required during the translunar voyage. The crew provided 72 minutes of color TV footage of Earth as it receded behind them. The first lunar orbit had an apolune of 196 miles and a perilune of 69 miles. When nearly circularized, the orbit was close to 69 miles above the lunar surface. During translunar flight to and from the Moon, as well as orbital maneuvers around the Moon and the descent and ascent to the lunar surface, the astronauts were almost completely dependent on the guidance system developed by Dr. Draper's Instrumentation Laboratory at MIT. Velocity adjustments of a few miles per hour (mph) in the correct direction when traveling up to 25,000 mph were essential and truly remarkable.

The next phase of the Apollo 10 mission called for Tom Stafford and Gene Cernan to separate the Lunar Module, *Snoopy*, from the Command Module, *Charlie Brown*, using the control-system thrusters. The descent engines then lowered the speed so that the lander's altitude was reduced to nearly 9 miles, its lowest point in orbit. The crew had no difficulty identifying landmarks. As Stafford said, "It looks as though all you have to do is put your test wheel down and we're there. The craters look flat and smooth on the bottom. It should be real easy." After separation from the descent stage, the ascent stage went into a violent oscillation, which provoked some unprintable expletives. But Tom took over

---

6. *Astronautics and Aeronautics, 1968: Chronology of Science, Technology, and Policy* (Washington, DC: NASA SP-4014, 1970), p. 11.

7. Ibid., p. 62.

Figure 34. Astronaut Edwin E. "Buzz" Aldrin, Lunar Module pilot of the first lunar landing mission, poses for a photograph beside the deployed United States flag during an Apollo 11 extravehicular activity (EVA) on the lunar surface. (NASA Image Number AS11-40-5875, also available at http://grin.hq.nasa.gov/ ABSTRACTS/GPN-2001-000012.html)

manual control and achieved the proper attitude. Rendezvous and docking were achieved without incident. On the return leg, there was more live color TV footage of both Earth and the Moon. On the eighth day, *Charlie Brown* landed, precisely on schedule, 3 to 4 miles from the recovery ship, USS *Princeton*. On NBC's *Meet the Press*, Tom Paine said that if the July lunar landing succeeded, there would be enough hardware for nine additional flights (this was later reduced to seven). He went on to say that it would take those flights and many more before men really began to understand Earth's twin planet.[8]

Wednesday, 16 July, 1969; Apollo 11; Neil Armstrong, Buzz Aldrin, and Mike Collins

Apollo 11 was scheduled for liftoff from Cape Canaveral's Launch Complex 39, Pad A, at 9:32 a.m.

EDT. Gene, our son Joe, and I arrived at the Cape the afternoon before. Flying with us to and from the Cape was Alexander de Seversky, the noted aircraft designer. The Jetstar arrived in time for us to have dinner in Cocoa Beach with Jim Webb and President Johnson. The dinner was most cordial, with toasts for Jim Webb's leadership and President Johnson's unflagging support. The next day, liftoff occurred on schedule, and the flight proceeded in a sequence nearly identical to that of Apollo 10. On the quarter-million-mile journey to the Moon, there were four TV broadcasts, with the longest lasting 96 minutes. The transmission was of excellent color, resolution, and general quality. The live pictures showed the interiors of the Command Module, *Columbia*, and the Lunar Lander, *Eagle*. Viewers could observe Earth, the Moon, and the opening of the hatch between the spacecraft modules, as well as housekeeping and food preparation. The lunar orbit was circularized at 75.6 miles of altitude.

---

8. Ibid., p. 142.

I had returned to Washington, DC, after the launch and then returned to Mission Control in Houston for the landing. The *Eagle* was separated from *Columbia* over the far side of the Moon and descended to 9.9 miles from the surface, at which point powered descent commenced. The location was 4.6 miles downrange of the planned location, so the landing point was significantly shifted. The feeling was tense in the control and also behind the glass where the handful of guests were located. It was soon noted that the *Eagle* was headed for the center of a crater containing boulders measuring 5 to 10 feet in length. Consequently, Neil raced beyond the crater by hand-controlling attitude and making throttle adjustments with the engine. Neil could extend the flight by 60 seconds before fuel shortage would require an abort. The clock was closing on zero when dust and shadows appeared in the foreground and suddenly Neil announced, "Houston, Tranquility Base here—the *Eagle* has landed."

Mission Control replied, "Roger, Tranquility. We copy you on the ground; you got a bunch of guys about to turn blue. We're breathing again, thanks a lot." The time was 4:18 p.m. EDT, 20 July.

Two hours after landing, the crew requested a walk on the Moon right away rather than 4.5 hours later, as originally planned. Doc Draper, Jackie Cochran (the famous aviatrix), and I went out for a quick bite, returning just after the postlanding checks. Shortly thereafter, Neil opened the hatch and descended *Eagle*'s ladder. As at least one-fifth of the world watched, he reached the lunar surface while saying, "One small step for a man, one giant leap for mankind."

Back in May 1961, Bill Fleming's committee had this to say about what should happen when a person first stood on the Moon: "Very little study has gone into precisely what operations would take place on the Moon or how they would be executed." In the interim, an extensive list of experiments was assembled. Neil first checked the surface and found that his foot's indentation was only a fraction of an inch. The Lunar Lander only penetrated 3 to 4 inches, the descent engine had not formed a crater, and the engine bell was about 1 foot above the surface. Our Moon is a solid structure.

Neil next filmed Buzz's descent onto the Moon, and the two together unveiled a plaque while reading its inscription: "Here men from planet Earth first set foot on the Moon July 1969 A.D. We came in peace for all mankind." Neil placed a camera a distance from the lander to photograph liftoff while Buzz experimented with movement in the low gravity—walking, running, leaping, and making two-footed kangaroo hops. He said that his agility was better than expected. Next, he deployed a solar-wind composition experiment and, with Neil, planted a pole with a 3-by-5-foot American flag (see figure 34). After saluting the flag, they phoned President Nixon. The President said, "As you talk to us from the Sea of Tranquility, it inspires us to redouble our efforts to bring peace and tranquility to Earth." The astronauts saluted the President and said it was an honor to represent the United States and the world.

Bulk samples of the lunar surface were then collected, and seismic equipment and a laser reflector were deployed. The astronauts then took two bore samples and picked up 20 pounds of "discretely selected material." After further photography, the EVA was completed; the astronauts returned to the lander, closed the hatch, and enjoyed 7 hours of rest. Apollo 11 returned to Earth by the same route as Apollo 10. Landing occurred in the Pacific, 15 miles from the USS *Hornet*, with the President and Tom Paine heading the welcoming committee (see figure 35). However, handshakes were not possible. There was concern that lunar pathogens might infect Earth and that Earth microorganisms might contaminate the lunar samples. So two-way biological barriers were created, one to protect the lunar samples and the other to protect life here on Earth. The lunar rocks then were flown immediately to the Lunar Receiving Laboratory in Houston. The three astronauts could talk and wave to the President from their "mobile quarantine facility," but there were no pats on the back.[9]

An estimated one million people viewed the Apollo liftoff from the Florida coast, including over 3,000 accredited press and TV commentators, and their interest held through the lunar landing and return. Congratulatory messages were printed in newspapers around the world. Tom Wicker of the

9.   Ibid., pp. 209–227.

Figure 35. President Richard M. Nixon welcomes the Apollo 11 astronauts aboard the USS Hornet, *prime recovery ship for the historic Apollo 11 lunar landing mission, in the central Pacific recovery area. (Left to right) Neil A. Armstrong, commander; Michael Collins, Command Module pilot; and Buzz Aldrin, Lunar Module pilot, are confined to the Mobile Quarantine Facility (MQF).(NASA Image Number S69-21365, also available at* http://grin.hq.nasa.gov/ABSTRACTS/GPN-2001-000007.html)

New York Times wrote about the Apollo 11 launch in a laudatory article on 17 July 1969:

> One could hardly watch the magnificent spectacle of the liftoff, let alone contemplate the feats of human ingenuity that made it possible, as well as the courage and skill of the flyers, without some reflection upon the meaning of the event . . . . The temptation is strong to fall back upon lyricism. The poetry of the thing has yet to find its expression in any of the earnest, proficient, Americans who have ventured away from the Earth; yet, the stunning beauty of man's most marvelous creation, as it rose in its majesty toward the unknown, toward the future, could be matched only by the profound sense of having been present at an end to something and therefore necessarily at a beginning.[10]

Dignitaries such as Soviet Premier Kosygin and Great Britain's Queen Elizabeth sent their warmest congratulations to the President. It was a giant, worldwide love-fest with only a few discordant voices. Historian Arnold Toynbee issued his views: "If we are going to go on behaving on Earth as we have behaved here so far, then a landing on the moon will have to be written off as one more shocking misuse of mankind's slender surplus product."[11] Many overseas intellectuals concurred with Toynbee. Regret was expressed in Swedish newspapers that America's feats of discovery were not matched by efforts toward the tremendous task of eliminating starvation on Earth.

10. Ibid., p. 228.

11. Ibid., p. 22.

The lunar program was initiated by President Kennedy with no such lofty ideas and goals in mind. The United States had landed man on the Moon, and at the very same time, the unmanned Soviet Luna 15 had apparently landed in the Sea of Crises and was no longer transmitting. Presumably it had crash-landed. In 1961, in a special address before Congress, President Kennedy spoke these words: "Now it is time to take longer strides—time for a great new American enterprise—time for this nation to take a clearly leading role in space achievement."[12] There was no question in the eyes of the world. This goal was achieved in July 1969.

What happened to the ongoing manned space efforts in both the USSR and the United States is summarized in the next and final chapter.

---

12. "Special Message to Congress on Urgent National Needs," 25 May 1961, *Public Papers of the Presidents of the United States, John F. Kennedy, January 20–December 31, 1961* (Washington, DC: U.S. Government Printing Office, 1962).

Chapter 7:
THE AFTERMATH

After the astronauts were welcomed aboard the *Hornet* by President Nixon and Tom Paine, they were brought to Houston aboard their biological laboratory, checked for pathogens, released, and greeted by family, friends, associates in Houston, and the world at large. An elegant testimonial was tendered to them by the President in Los Angeles. All the place settings were transported from the White House. I flew out one day and back the next with the Joint Chiefs. I played more bridge in two days then I had since college. At the dinner, the dais was quite high, but I was just able to shake hands with Neil, Buzz, and Mike and congratulate them on a job extremely well done. Of course, the mission wasn't complete until the doctors were finished examining the astronauts. The engineers analyzed the glitches, such as the computer overload on landing. (The rendezvous radar was accidentally left on, thereby feeding excessive signals to the computer.) And scientists were provided with the sample returns and other lunar data (130 laboratories, one-third of them overseas, participated in the analysis of the lunar specimens).

Was There a Space Race?

Was there truly a space competition between the Soviet Union and the United States? The accompanying chart (figure 36) shows the manned launches during the period from 1961 through 1970. The three Soviet spacecraft on the left start with Vostok, then Voskhod, and finally Soyuz, the latter still in use today. On the right side are the U.S. manned capsules, Mercury, Gemini, and Apollo. To gain a mental picture of the competition, note that the first orbital mission, for example, was flown by Vostok, followed almost a year later by Mercury. Up until the end of 1968, the Soviets were in the lead with six firsts to the United States' one.

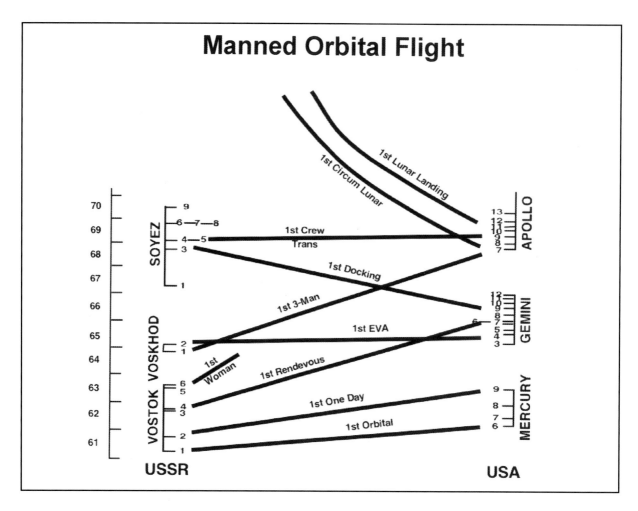

Manned Orbital Flight

Figure 36. Comparison of the Soviet and U.S. manned launches during the period from 1961 through 1970. (Source: Robert C. Seamans, Jr., papers, MC 247, Institute Archives and Special Collections, MIT Libraries, Cambridge, MA.)

But were the Soviets planning to land on the Moon? You bet! At the time of our lunar landing, the United States had been tracking the USSR N-1 booster development with overhead satellite photography for several years. The photos first showed a large building under construction with rails in and out. The rails came from their industrial area and led to an emerging large launchpad. Sometime later, the photos caught sight of the vehicle proceeding horizontally to the launchpad. The Soviets always waited to erect their vehicles until they reached the pad. The booster was never given a name, just the designation N-1. Figure 37 shows the N-1 on the pad with its umbilical tower, along with the umbilical arms that provided ready access. The proportions of N-1 can be seen in figure 38 in comparison with those of the Saturn V. The two vehicles are of comparable height and weight. A vehicle of this size could be used to launch a space station, as the United States did in 1973 with Skylab, but the most likely purpose was for lunar exploration.

It's my understanding that the Russians planned to explore the Moon with the lander and then rendezvous and dock with the Soyuz in a maneuver similar to the crew transfer practiced by Soyuz 3 and 4 in mid-January 1969. A rendezvous also would have occurred earlier, in Earth orbit. The N-1 would ferry the Earth-escape rockets, lunar propulsion systems, and lunar lander into orbit, followed by the cosmonauts aboard the Soyuz, who would rendezvous and dock with their lunar chariot.

The plans for a lunar landing were denied by the Russians until the fall of the Soviet Union. During that uncertain period, three professors from MIT's Department of Aeronautics and Astronautics were visiting aerospace facilities in Moscow in 1989. As they were passing the doorway into a laboratory, they spotted interesting hardware. "Can we go in?" they asked. There was a shrug, so in they went, and they soon spotted a capsule mounted on top of a bell-shaped configuration. A sign in Russian said

Figure 37. N-1 on the pad with its umbilical tower, along with the umbilical arms that provided ready access. (Available at http://grin.hq.nasa.gov/ABSTRACTS/GPN-2002-000188.html)

circle of senior government officers, and key military and launch personnel at the cosmodrome. The explosion just after liftoff was devastating (figure 40). Havoc was wreaked over a wide area. A Russian student at a seminar I was conducting at MIT volunteered that his father had been in an engineering building several miles away at the time and had been thrown to the floor amidst a multitude of glass shards.

But there was still one more move on the chessboard. All systems were "go" for an unmanned sample return vehicle, Luna 15 to be launched at the same time as Apollo 11. The two vehicles actually orbited the Moon at the same time, but the Soviets were uncertain about the landing topography. The descent of Luna 15 was delayed 18 hours before it received approval to descend; then Luna 15 was to land 6 minutes after Armstrong and Aldrin left the Moon. However, with several minutes left before landing, all signals from Luna 15 abruptly terminated. Later, it was determined that the vehicle hit a mountain peak. The Soviets' checkmate would have occurred if the U.S. astronauts had been unsuccessful and the Soviets had collected a lunar sample. How foolhardy the United States would have appeared, with the Soviets displaying their Moon rocks, if the astronauts had been killed or left forever in space. In actuality, the news release from Tass (the Soviet news agency) said, "Luna 15's record program had been completed and the spacecraft had reached the moon in the preset area."[1]

Our Knowledge of the Soviet Space Program

As was said in the report to the Vice President signed by McNamara and Webb on 8 May 1961, "Our cards are and will remain up, and theirs are face-down." How prophetic was this statement! Our intelligence was severely limited. We knew that there was a single individual directing the Soviets' space effort. Occasionally, our intelligence included fragmentary conversations intercepted when he called his office from his car, but to my knowledge, no substantive information was received from these intercepts. Not until after his death did his name and background become common knowledge. At that time, he was honored and

"lunar lander," which was recognized by the Americans. "Can we photograph?" Another shrug. The photo taken by Jack Kerrebrock (figure 39) shows the ascent capsule with the lander underneath, with Professor Larry Young in front, examining it and quizzing the professors' Russian counterparts.

The Soviets played for a checkmate up until the safe return of Apollo 11 with Armstrong, Aldrin, and Collins aboard. Even with Frank Borman celebrating the Fourth of July at the U.S. embassy in Moscow, the attempt to launch the N-1 the night of 3 July was completely unknown except to a small

---

1. *Astronautics and Aeronautics, 1969: Chronology of Science, Technology, and Policy* (Washington, DC: NASA SP-4014, 1970), p. 236.

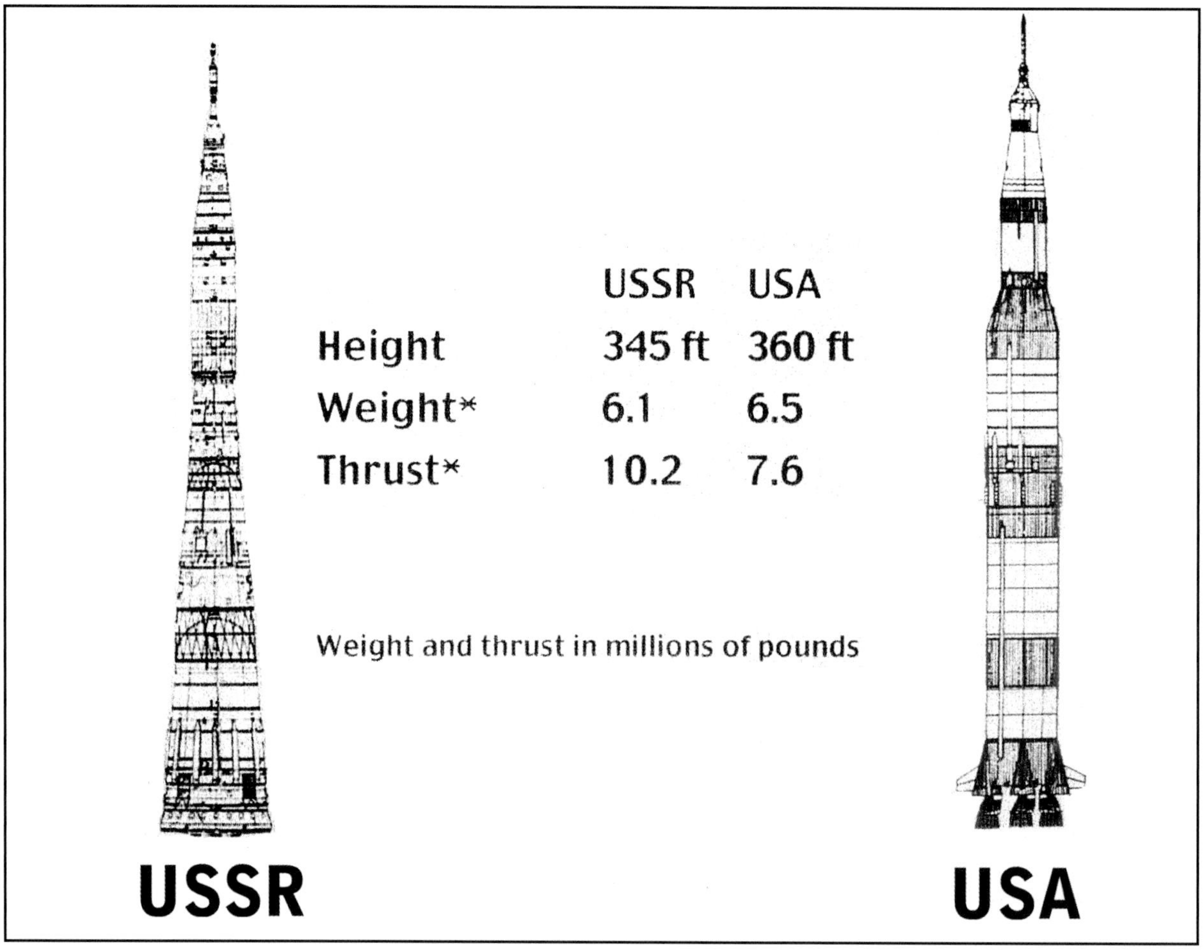

	USSR	USA
Height	345 ft	360 ft
Weight*	6.1	6.5
Thrust*	10.2	7.6

Weight and thrust in millions of pounds

USSR

USA

Figure 38. Comparison of the Soviet N-1 with the U.S. Saturn V. (Source: Robert C. Seamans, Jr., papers, MC 247, Institute Archives and Special Collections, MIT Libraries, Cambridge, MA.)

buried in the Kremlin Wall. Our only solid Central Intelligence Agency (CIA) information came from the Corona satellite photography. We observed the construction of the Soviets' large launch vehicle assembly building with rails leading to an underdeveloped area. Later, a launch area materialized, and on one fortuitous occasion, a large vehicle was spotted ready for the launchpad. After the fact, photographs showed the devastation from an explosion on the pad and then a second soon after liftoff. But we had no direct evidence of their lunar program until the visit of three MIT faculty to their Moscow Aviation Institute space laboratory in December 1989 (see figure 39).

However, the Soviets publicly announced each of their flights after the fact. So we knew the names of the cosmonauts, dates, and stated results.

We checked the veracity of their releases against the dates and ephemera (apogee, perigee, inclination) of space objects stored by the Air Force in Cheyenne Mountain, Colorado. The Air Force was tracking about 4,500 orbital objects.[2] NASA would be informed when a new object came over the horizon, as well as if and when it left orbit and reentered the atmosphere. In this way, the existence of each Soviet mission could be verified, but the intent was a matter of speculation. The inclusion in the text of each Soviet flight in the correct chronological order provides the same understanding of Soviet plans that was available to NASA. For example, as has already been discussed, the Soviet launch of two manned spacecraft a day apart and the subsequent rendezvous could be construed as either part of a lunar or a military maneuver. However, when these things were coupled with the crew

_____

2. Dwayne A. Day, "The Secret at Complex J," *Air Force Magazine* (July 2004): 72.

Figure 39. A photograph of the Soviet Lunar Lander and Return Vehicle taken at the Moscow Aviation Institute on 28 November 1989. The occasion was a visit by three Massachusetts Institute of Technology (MIT) professors—Lawrence Young (left center), Jack Kerrebrock (the photographer), and Edward Crawley (not pictured).

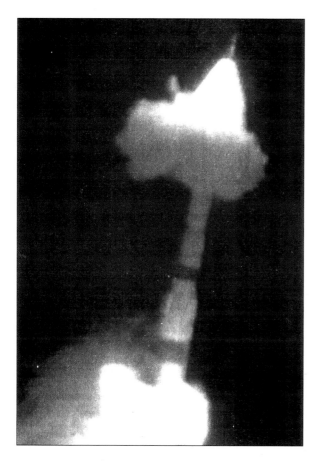

Figure 40. Soviet disaster: the N-1 explodes

transfer in later flights and the appearance of a large booster, it became clear to NASA that the Soviets were preparing for manned lunar missions.

Why the Soviet Success and Then Disaster?

Why the Soviets failed is still a matter of speculation. I would list the following as possible reasons:

Sergey Korolev died on the operating table with a burst appendix in January 1966. His stature placed him in the Kremlin Wall with numerous honors. He was under wraps throughout his gulag years and the period when he directed the Soviet space effort. He was respected for his consummate skill and feared because of his relationship with Khrushchev and other high-ranking officials. He knew how to run their space program, and he had the chits to achieve his goals. The prevailing view of their fire drill in July 1969 after his death is that it was a desperate gamble.

The Soviets were lacking the technology available to the United States. They had started the '60s clearly ahead, with boilerplate ballistic missiles capable of lifting large payloads into orbit. Korolev played his cards extremely well, but by the end of the decade, the Soviets had no high-impulse rockets for their upper stages. Thanks to wise decisions by Abe Silverstein in Keith Glennan's era, oxygen-hydrogen engines were available for Saturn I and Saturn V. To compensate, the Soviets required one-third more thrust at liftoff, and they didn't have a large engine like the F-1. For that reason, their first stage had to harness 30 rocket motors.

Moreover, the silicon chip was just beginning to revolutionize the electronics industry in the United States. At the Draper Laboratory, for example, the question arose whether guidance and control for Apollo were to be restricted to printed circuitry. Doc Draper wanted to take advantage of Texas Instruments' latest developments, and he did.

During the assembly of the Apollo/Saturn in the Vertical Assembly Building, the instruments on the vehicles read out to monitors in Launch Control just as they did during checkout and prior to launch. Then, in the final 2 minutes prior to liftoff, the signals were automatically sequenced. If any reading was out of tolerance, there would be a hold and the fault would be investigated. Finally, the United States would never have succeeded with Apollo if we'd plodded through 20 to 30 flights of Saturn I and Saturn V to achieve man rating. The funding wouldn't have been available. All-up systems testing was essential, and the checkout capability was a strong contributor to NASA's success.

What Was Next for Apollo?

After Apollo 11, there were eight more lunar voyages planned and funded, but Apollo 13's lunar landing was scrubbed because of a severe power loss in transit to the Moon. The crew's return was the greatest display of heroics and skill of any NASA mission. In addition, the Nixon administration canceled two missions for cost reasons. The remaining five successful lunar landings conducted important studies aided, in the final three missions, by a lunar buggy (see figure 41) with a 15-mile roving radius. This flexibility, plus the increased precision of the landings, led to important discoveries, including the finding of material dating from near

Figure 41. The Lunar Rover provided the astronauts with an opportunity to explore the landing area to distances of 10 miles. This capability was available for the final three lunar missions. (NASA Image Number AS17-147-22526, also available at http://grin.hq. nasa.gov/ABSTRACTS/GPN-2000-001139.html*)*

the time of the Earth-Moon marriage four billion years ago.

There were two other uses of the Apollo hardware: Skylab and Apollo-Soyuz. The third stage of Saturn V became Skylab: instead of fuel, oxidizer, and rocket engines, the stage carried a habitat, a research laboratory, and extensive solar paddles for power. Placed in orbit by the first two stages of the Saturn V, Skylab was 118 feet long and 22 feet in diameter, and it weighed about 200,000 pounds. When Skylab arrived in orbit on 14 May 1973, it required intensive care. The temperature awning was torn off, and some of the solar panels had been badly damaged; others had fallen off.[3] While ground con-

trol kept changing Skylab's attitude to minimize solar heating, the crew of Charles Conrad, Paul Weitz, and Joseph Kerwin prepared for its resuscitation. On 25 May 1973, they rendezvoused with Skylab and performed miracles. A cable was cut to permit the existing solar panels to deploy. Then, with sufficient power, they entered the extrawarm cabin and deployed a large parasol through a hole in the craft's skin. When it was opened, necessary solar shade was provided. The crew stayed aboard for 28 days. The second crew of Alan Bean, Jack Lousma, and Owen Garriott vacationed aboard for 59.5 days, arriving on 28 July. Gerald Carr, William Pogue, and Edward Gibson had an 84-day visit from 16 November 1973 to 8 February 1974.

---

3. *Astronautics and Aeronautics, 1969: Chronology of Science, Technology, and Policy* (Washington, DC: NASA SP-4018, 1975), pp. 142–152.

Inside the laboratory, 270 scientific and technical investigations covering the fields of space physics, stellar and galactic astronomy, solar physics, bioscience, and space medicine were planned. About a hundred principal investigators participated. The solar observations were conducted with the aid of seven solar telescopes. Six of these recorded on film, and the seventh relied on the transmission of photoelectric data. These telescopes made available 195,000 exposures of the Sun and its corona to land-based astronomers. By any measure, Skylab was a great success. If the Saturn I had not been canceled, Skylab could have been more permanent, perhaps serving as the focal point for the present space laboratory. Unfortunately, in 1979, approximately a year prior to the beginning of Shuttle operations, the atmosphere's upper reaches slowed Skylab until it fell to Earth in a fiery ball.

There was still one more Apollo mission—the linking of Apollo and Soyuz 19 in July of 1975. This primarily political venture was agreed to by President Nixon and Soviet Premier Aleksei Kosygin in 1972. Planning and equipping the Apollo and Soyuz for the mission took visits to the Cosmodrome in Russia and the Manned Spacecraft Center in Houston. The United States constructed the docking module, which permitted the astronauts and cosmonauts to meet directly, without extravehicular activity. The two-man Soyuz was launched approximately 7.5 hours ahead of Tom Stafford, Deke Slayton, and Vance Brand in Apollo. Rendezvous and docking took place after the 29th orbit of Soyuz.[4] When Tom Stafford greeted the cosmonauts, he did so in Russian. One of the Russians muttered, "I wish I'd thought of doing that." (Tom had been taking Russian lessons for nearly a year.) After two days of joint activities and three to complete a list of experiments, Apollo returned to Earth, and the Apollo days were over.

Final Thoughts

Following the Apollo era, I would call Patsy Webb, Jim Webb's wife, and ask about his health and arrange a visit, if possible, whenever I returned to Washington. He fought Parkinson's with courage. Sometimes he would be wheeled in by an attendant; sometimes he would be on his electric scooter. Often he had a patch over one eye to avoid seeing double. Whether it was the Dopamine he took ahead of time to gain strength in his body or his interest in the conversations, I never knew—but invariably, he'd walk me to the door when I left. Sometimes he'd give me an article or a book to read, and he'd ask for my evaluation of the contents. Other times, he'd probe me with questions like, "What do you hope to accomplish before you kick the bucket?" But although he never asked me why I had resigned from NASA in 1967, I think he knew and agreed. We had become incompatible, and our mutual trust of six years had disappeared. Pages 143 to 150 of my autobiography *Aiming at Targets* discuss my relationship with Jim after the fire.

Often he'd ask the question, "We thought we were building a space capability for years to come. Why didn't it happen?" Part of the answer was the cost. I once heard President Johnson's Director of Management and Budget being interviewed on public radio. He was explaining the budget during President Johnson's final two years in office. The country was mired in Vietnam, and the Great Society was getting in gear, both with attendant increased costs. And in addition, the President refused to increase taxes. So where did he look for relief? Obviously, the space program. At first he tried to talk the President into a cancellation of the lunar objective. Johnson said he owed it to President Kennedy to complete. But the Soviets appeared out of the race, so why not wait a few years, he argued. Johnson insisted that the lunar landing occur within the decade. There was still one moderate-size expenditure for the Office of Management and Budget, OMB, to strike, namely, Apollo Applications. Starting in the mid-sixties, considerable thought and effort went into future planning under this rubric. George Mueller had the action, and he had selected E. Z. Grey to mount the studies. The Apollo/Saturn capability could have been used separately or in tandem for a wide variety of missions including large orbiting spacecraft for geophysics and astronomy, a permanent space station, a modest base on the Moon, and large unmanned payloads to the planets.

At the same time that Congress was pressing us for our plans, the OMB was picking our pockets for the benefit of other national goals. President Nixon

4. Ibid., pp. 131–137.

finally supported the concept of reusability in the form of the Shuttle. The five approved vehicles were to be the basis for 500 missions. After 20 years, 133 missions have been flown, but only three Shuttles remain, and the International Space Station is only one-third complete. Over the years, bold new ventures have been suggested, including manned bases on the Moon and manned landing on Mars. There are many unfulfilled dreams of such missions still in the minds of those who participated in Apollo; however, the Saturn V would be extremely costly to resurrect. The Shuttle cannot carry large enough payloads for many of President George W. Bush's initiatives. The Shuttle is currently reserved for the Space Station and will be retired from service when the Station is complete.

A National Vision for Space Exploration

During the time I was organizing my thoughts on the Apollo Program, a 40-year-old relic of the past, President Bush announced the Vision for Space Exploration, a space program 20 to 30 years into the future. His goal includes sustained and affordable human and robotic projects to explore the solar system and beyond. Human presence would start with a return to the Moon as early as 2015 but no later than 2020. The lunar missions can have immediate significance, for example, to serve as astronomical outposts, and they are also part of the preparation for human exploration of Mars and other destinations. Judging from the past, the technology and our national needs and objectives will be altered before the vision's completion. So the project is difficult to appraise because not only is it in its infancy, but the first step hasn't been authorized by Congress yet.

So what can be said at this time? The overall concept makes sense. The Vision is considered a journey, not a destination. The attempt will be made to develop modules such as the Crew Exploration Vehicle, which can be used on many types of missions. Not much has been revealed about the launch vehicle for Earth escape. Is a large vehicle contemplated, or will a series of launches followed by rendezvous and docking put the mission in play? Nuclear propulsion and power are planned for the longer journeys. The high specific impulse that can be provided by nuclear propulsion is desirable for long distances. The time for long-duration travel can be halved (most important for manned flight), but will the use of nuclear fuel be acceptable to the U.S. public and the world community?

The plan wisely encompasses both manned and robotic missions. Thorough investigation of lunar and planetary pathways by robots must be a prerequisite to manned excursions. And long-duration stay times on the lunar and Martian surfaces by robotic missions must precede man's adventure to these distant locales.

So the concept of the Vision for Space Exploration appears sound, but what will be learned and what will be gained? And how will the Vision be managed? A competent team is now reviewing and planning the future, but will they be able to overcome the vicissitudes of changing political agendas?

At the time of Apollo, NASA had a 10-year plan that was updated annually for Congress. Perhaps President Bush's space journey can be viewed as a 30-year plan made up of a series of defined objectives important in their own right. Then the congressional approvals could be directed toward short-term objectives in a long-term framework. Perhaps more stable budget requirements would result.

Might some new approach or technology be conceived to greatly reduce the cost? Perhaps President Bush's initiative will trigger such an enabling concept. Certainly man can be remarkably creative, and if true needs arise, I believe man will find solutions.

Appendix 1

Transmittal Letter with Report Attached from NASA Administrator James E. Webb to President John F. Kennedy, 23 March 1961

George Low, chief of Manned Space Flight, conducted a manned space study during the last two weeks of the Eisenhower administration and the early weeks of the Kennedy administration. The results are summarized in figure 1, which shows that with increased funding, three manned crews could orbit Earth in 1965; circle the Moon in 1967, using Saturn vehicles; and land on the Moon as early as 1970, using a Nova vehicle.

In order to conduct these missions at these earlier dates, increased annual funding was required, although the total funding might remain the same. I associated the increased budget request for fiscal year 1962 with the earlier flight dates recommended by George Low when I ad-libbed my summary at the end of our meeting with the President. He was well aware of the USSR's tremendous advantage in weight-lifting capability, so my summary hit a responsive chord. He liked the summary and asked for it in writing the following day. I put pencil to paper that evening. The typed result, along with Mr. Webb's letter of transmittal, was sent to the President on 23 March 1961, the day after our meeting.

March 23, 1961

MEMORANDUM for the President

The attached memorandum prepared by our Associate Administrator, Dr. Seamans, responds to your request of yesterday that he furnish you and the Vice President with certain information concerning NASA's plans.

Original signed by

James E. Webb

Administrator

Attachment

March 23, 1961

MEMORANDUM

To: The Administrator

From: Associate Administrator

Subject: Recommended increases in FY62 Funding for Launch Vehicles and Manned Space Exploration.

The funding rates of five projects were discussed at the NASA-BOB conference with the Vice President and the President on March 22, 1961. An agenda prepared prior to the meeting summarized the objectives of these projects and indicated in each case the effect of the funding rate on the schedule. The projects are listed below along with a tabulation of the current and recommended funding rates for FY1962.

Project	Current Funding Rate	Recommended New Funding Rate	Net Change
Centaur	$53.9	$80.9	+$27.0
Saturn C-2	$20.0	$98.0	+$78.0
Prototype Engine for Nuclear Rocket	$13.5	$41.0	+$27.5
Nova Type (F-1) Engine	$33.4	$43.7	+$10.3
Multi-manned Orbital Laboratory	$29.5	$77.2	+$47.7
	$150.3	$340.8	$190.5

The FY 1962 increase in these key areas as discussed by NASA would amount to $190.5 million out of the total increase of $303.6 million proposed as a revised NASA budget.

The multi-manned orbital laboratory is contingent upon the Saturn C-1 which is adequately funded, and a new spacecraft for which NASA recommends an increase from $29.5 to $77.2. This increase starts an accelerated program leading to multi-manned orbital flights in 1965 rather than 1967.

The multi-manned circumlunar flight requires the Saturn C-2 and a spacecraft which will evolve from the design of the orbital spacecraft. The recommended $73 million increase in FY1962 funding for the Saturn C-2 leads to the completion of the Saturn development in 1966, and manned circumlunar flight in 1967 rather than in 1969.

A manned lunar landing requires a new launch vehicle with capabilities beyond Saturn. This vehicle, called NOVA, is still under study. It would use a first-stage cluster of the 1.5 million pound thrust, chemically fueled engines which we have under development. We are requesting $10.3 million additional over the present FY 1962 budget to accelerate the engine development. The first manned lunar landings depend upon this chemical engine as well as on the orbital and circumlunar programs and can be achieved in 1970 rather than 1973.

Subsequent lunar base operations or manned planetary explorations depend upon having a nuclear rocket to provide the much heavier payloads required for such missions. We recommend a FY 1962 increase for the development of a prototype flight nuclear engine. An acceleration of $27.5 million in NASA funds matched by an AEC increase of $17.0 million will permit initial flight tests in early 1967 instead of 1968. Further development of this type engine for use in an upper stage of the Nova will provide a payload weight capability nearly double that of an all chemically-fueled vehicle.

Increase to the level now proposed for the Centaur, Saturn, large chemical-engine, nuclear engine, and multi-manned spacecraft will increase the rate of closure on the USSR's lead in weight lifting capability and significantly advance our manned exploration of space beyond Project Mercury.

Robert C. Seamans, Jr.
Associate Administrator

Appendix 2

James E. Webb's Letter to President Kennedy of 30 November 1962, Requested by the President at Our Meeting on 21 November 1962

The discussion with President Kennedy on 1 November revolved around the issue of a $400-million supplemental request for fiscal year 1963. Brainerd Holmes recommended the supplemental as a means for advancing the lunar landing date from 1967 to 1966. Mr. Webb, Dr. Dryden, and I were strongly opposed. In 1961, we had gained approval from Congress for an FY 1962 budget increase from $1.1 billion to $1.8 billion, and Congress had appropriated $3.7 billion for FY 1963. In our view, Congress would balk at a still further increase, and we didn't feel that NASA could efficiently sustain still further growth.

At the meeting, the President championed the possibility of the earlier lunar landing. When he understood the political consequence of the supplemental, he pressed hard for a reprogramming of funds from nonlunar missions. The debate that ensued centered on this issue. The President argued that the manned lunar landing was one of the two highest priority nondefense projects of his administration. He felt that other efforts at NASA were useful but could be delayed. Jim Webb argued that many of the scientific and technical programs, although not directly managed by Brainerd Holmes, provided essential design information for the manned lunar landing. He also noted that other programs were important in their own right. Some were time-sensitive, some were joint efforts with other nations, and some were related to DOD and other government agencies.

So at first, President Kennedy argued that the manned lunar landing was the highest priority of NASA's missions, and Mr. Webb argued that NASA's goal was preeminence in space. As the meeting proceeded, the President conceded that there might be scientific and technical efforts providing essential data for the lunar mission, and Mr. Webb conceded only that NASA was already proceeding at flank speed and couldn't accelerate the lunar mission further. At the meeting's end, the President said, "Maybe we're not too far apart; write me a summary of your views on NASA's priorities." The extensive letter responding to the President's request summarizes NASA's manned lunar effort, discusses related and unrelated activities, and contains a bit of NASA's fundamental creed. For example, in the section "Advanced Research and Technology," the last sentence in the first paragraph reads, "The philosophy of providing for an intellectual activity of research and an interlocking cycle of application must be a cornerstone of our National Space Program."

The letter achieved its purpose. There was no further discussion of supplementals and reprogramming to achieve a lunar landing at an earlier date. Most important, "preeminence in space" became NASA's watchword.

NATIONAL AERONAUTICS AND SPACE ADMINISTRATION

WASHINGTON 25, D.C.

OFFICE OF THE ADMINISTRATOR

November 30, 1962

The President

The White House

Dear Mr. President:

At the close of our meeting on November 21, concerning possible acceleration of the manned lunar landing program, you requested that I describe for you the priority of this program in our over-all civilian space effort. This letter has been prepared by Dr. Dryden, Dr. Seamans, and myself to express our views on this vital question.

The objective of our national space program is to become pre-eminent in all important aspects of this endeavor and to conduct the program in such a manner that our emerging scientific, technological, and operational competence in space is clearly evident.

To be pre-eminent in space, we must conduct scientific investigations on a broad front. We must concurrently investigate geophysical phenomena about the earth, analyze the sun's radiation and its effect on earth, explore the moon and the planets, make measurements in interplanetary space, and conduct astronomical measurements.

To be pre-eminent in space, we must also have an advancing technology that permits increasingly large payloads to orbit the earth and to travel to the moon and the planets. We must substantially improve our propulsion capabilities, must provide methods for delivering large amounts of internal power, must develop instruments and life support systems that operate for extended periods, and must learn to transmit large quantities of data over long distances.

To be pre-eminent in operations in space, we must be able to launch our vehicles at prescribed times. We must develop the capability to place payloads in exact orbits. We must maneuver in space and rendezvous with cooperative spacecraft and, for knowledge of the military potentials, with uncooperative spacecraft. We must develop techniques for landing on the moon and the planets, and for re-entry into the earth's atmosphere at increasingly high velocities. Finally, we must learn the process of fabrication, inspection, assembly, and check-out that will provide vehicles with life expectancies in space measured in years rather than months. Improved reliability is required for astronaut safety, long duration scientific measurements, and for economical meteorological and communications systems.

In order to carry out this program, we must continually up-rate the competence of Government research and flight centers, industry, and universities, to implement their special assignments and to work together effectively toward common goals. We also must have effective working relationships with many foreign countries in order to track and acquire data from our space vehicles and to carry out research projects of mutual interest and to utilize satellites for weather forecasting and world-wide communications.

Manned Lunar Landing Program

NASA has many flight missions, each directed toward an important aspect of our national objective. The manned lunar landing program requires for its successful completion many, though not all, of these flight missions. Consequently, the manned lunar landing program provides currently a natural focus for the development of national capability in space and, in addition, will provide a clear demonstration to the world of our accomplishments in space. The program is the largest single effort within NASA, constituting three-fourths of our budget, and is being executed with the utmost urgency. All major activities of NASA, both in headquarters and in the field, are involved in this effort, either partially or full time.

In order to reach the moon, we are developing a launch vehicle with a payload capability 85 times that of the present Atlas booster. We are developing flexible manned spacecraft capable of sustaining a crew of three for periods up to 14 days. Technology is being advanced in the areas of guidance and navigation, re-entry, life support, and structures—in short, almost all elements of booster and spacecraft technology.

The lunar program is an extrapolation of our Mercury experience. The Gemini spacecraft will provide the answers to many important technological problems before the first Apollo flights. The Apollo program will commence with earth orbital maneuvers and culminate with the one-week trip to and from the lunar surface. For the next five to six years there will be many significant events by which the world will judge the competence of the United States in space.

The many diverse elements of the program are now being scheduled in the proper sequence to achieve this objective and to emphasize the major milestones as we pass them. For the years ahead, each of these tasks must be carried out on a priority basis.

Although the manned lunar landing requires major scientific and technological effort, it does not encompass all space science and technology, nor does it provide funds to support direct applications in meteorological and communications systems. Also, university research and many of our international projects are not phased with the manned lunar program, although they are extremely important to our future competence and posture in the world community.

Space Science

As already indicated, space science includes the following distinct areas: geophysics, solar physics, lunar and planetary science, interplanetary science, astronomy, and space biosciences.

At present, by comparison with the published information from the Soviet Union, the United States clearly leads in geophysics, solar physics, and interplanetary science. Even here, however, it must be recognized that the Russians have within the past year launched a major series of geophysical satellites, the results of which could materially alter the balance. In astronomy, we are in a period of preparation for significant advances, using the Orbiting Astronomical Observatory which is now under development. It is not known how far the Russian plans have progressed in this important area. In space biosciences and lunar and planetary science, the Russians enjoy a definite lead at the present time. It is therefore essential that we push forward with our own programs in each of these important scientific areas in order to retrieve or maintain our lead, and to be able to identify those areas, unknown at this time, where an added push can make a significant breakthrough.

A broad-based space science program provides necessary support to the achievement of manned space flight leading to lunar landing. The successful launch and recovery of manned orbiting spacecraft in Project Mercury depended on knowledge of the pressure, temperature, density, and composition of the high atmosphere obtained from the nation's previous scientific rocket and satellite program. Considerably more space science data are required for the Gemini and Apollo projects. At higher altitudes than Mercury, the spacecraft will approach the radiation belt through which man will travel to reach the moon. Intense radiation in this belt is a major hazard to the crew. Information on the radiation belt will determine the shielding requirements and the parking orbit that must be used on the way to the moon.

Once outside the radiation belt, on a flight to the moon, a manned spacecraft will be exposed to bursts of high speed protons released from time to time from flares on the sun. These bursts do not penetrate below the radiation belt because they are deflected by the earth's magnetic field, but they are highly dangerous to man in interplanetary space.

The approach and safe landing of manned spacecraft on the moon will depend on more precise information on lunar gravity and topography. In addition, knowledge of the bearing strength and roughness of the landing site is of crucial importance, lest the landing module topple or sink into the lunar surface.

Many of the data required for support of the manned lunar landing effort have already been obtained, but as indicated above there are many crucial pieces of information still unknown. It is unfortunate that the scientific program of the past decade was not sufficiently broad and vigorous to have provided us with most of these data. We can learn a lesson from this situation, however, and proceed now with a vigorous and broad scientific program not only to provide vital support to the manned lunar landing, but also to cover our future requirements for the continued development of manned flight in space, for the further exploration of space, and for future applications of space knowledge and technology to practical uses.

Advanced Research and Technology

The history of modern technology has clearly shown that pre-eminence in a given field of endeavor requires a balance between major projects which apply the technology, on the one hand, and research which sustains it on the other. The major projects owe their support and continuing progress to the intellectual activities of the sustaining research. These intellectual activities in turn derive fresh vigor and motivation from the projects. The philosophy of providing for an intellectual activity of research and an interlocking cycle of application must be a cornerstone of our National Space Program.

The research and technology information which was established by the NASA and its predecessor, the NACA, has formed the foundation for this nation's pre-eminence in aeronautics, as exemplified by our military weapons systems, our world market in civil jet airliners, and the unmatched manned flight within the atmosphere represented by the X-l5. More recently, research effort of this type has brought the TFX concept to fruition and similar work will lead to a supersonic transport which will enter a highly competitive world market. The concept and design of these vehicles and their related propulsion, controls, and structures were based on basic and applied research accomplished years ahead. Government research laboratories, universities, and industrial research organizations were necessarily brought to bear over a period of many years prior to the appearance before the public of actual devices or equipment.

These same research and technological manpower and laboratory resources of the nation have formed a basis for the U.S. thrust toward pre-eminence in space during the last four years. The launch vehicles, spacecraft, and associated systems including rocket engines, reaction control systems, onboard power generation, instrumentation and equipment for communications, television and the measurement of the space environment itself have been possible in this time period only because of past research and technological effort. Project Mercury could not have moved as rapidly or as successfully without the information provided by years of NACA and later NASA research in providing a base of technology for safe re-entry heat shields, practical control mechanisms, and life support systems.

It is clear that a pre-eminence in space in the future is dependent upon an advanced research and technology program which harnesses the nation's intellectual and inventive genius and directs it along selective paths. It is clear that we cannot afford to develop hardware for every approach but rather that we must select approaches that show the greatest promise of payoff toward the objectives of our nation's space goals. Our research on environmental effects is strongly focused on the meteoroid problem in order to provide information for the design of structures that will insure their integrity through space missions. Our research program on materials must concentrate on those materials that not only provide meteoroid protection but also may withstand the extremely high temperatures which exist during re-entry as well as the extremely low temperatures of cryogenic fuels within the vehicle structure. Our research program in propulsion must explore the concepts of nuclear propulsion for early 1970 applications and the even more advanced electrical propulsion systems that may become operational in the mid-1970's. A high degree of selectivity must be and is exercised in all areas of research and advanced technology to ensure that we are working on the major items that contribute to the nation's goals that make up an over-all pre-eminence in space exploration. Research and technology must precede and pace these established goals or a stagnation of progress in space will inevitably result.

Space Applications

The manned lunar landing program does not include our satellite applications activities. There are two such program areas under way and supported separately: meteorological satellites and communications satellites. The meteorological satellite program has developed the TIROS system, which has already successfully orbited six spacecraft and which has provided the foundation for the joint NASA-Weather Bureau planning for the national operational meteorological satellite system. This system will center on the use of the Nimbus satellite which is presently under development, with an initial research and development flight expected at the end of 1963. The meteorological satellite developments have formed an important position for this nation in international discussions of peaceful uses of space technology for world benefits.

NASA has under way a research and development effort directed toward the early realization of a practical communication satellite system. In this area, NASA is working with the Department of Defense on the Syncom (stationary, 24-hour orbit, communications satellite) project in which the Department of Defense is providing ground station support for NASA's spacecraft development; and with commercial interests, for example, AT&T on the Telstar project. The recent "Communications Satellite Act of 1962" makes NASA responsible for advice to and cooperation with the new Communications Satellite Corporation, as well as for launching operations for the research and/or operational needs of the Corporation. The details of such procedures will have to be defined after the establishment of the Corporation. It is clear, however, that this tremendously important application of space technology will be dependent on NASA's support for early development and implementation.

University Participation

In our space program, the university is the principal institution devoted to and designed for the production, extension, and communication of new scientific and technical knowledge. In doing its job, the university intimately relates the training of people to the knowledge acquisition process of research. Further, they are the only institutions which produce more trained people. Thus, not only do they yield fundamental knowledge, but they are the sources of the scientific and technical manpower needed generally for NASA to meet its program objectives.

In addition to the direct support of the space program and the training of new technical and scientific personnel, the university is uniquely qualified to bring to bear the thinking of multidisciplinary groups on the present-day problems of economic, political, and social growth. In this regard, NASA is encouraging the universities to work with local industrial, labor, and governmental leaders to develop ways and means through which the tools developed in the space program can also be utilized by the local leaders in working on their own growth problems. This program is in its infancy, but offers great promise in the working out of new ways through which economic growth can be generated by the spin-off from our space and related research and technology.

International Activity

The National Space Program also serves as the base for international projects of significant technical and political value. The peaceful purposes of these projects have been of importance in opening the way for overseas tracking and data acquisition sites necessary for manned flight and other programs which, in many cases, would otherwise have been unobtainable. Geographic areas of special scientific significance have been opened to cooperative sounding rocket ventures of immediate technical value. These programs have opened channels for the introduction of new instrumentation and experiments reflecting the special competence and talent of foreign scientists. The cooperation of other countries—indispensable to the ultimate achievement of communication satellite systems and the allocation of needed radio frequencies—has been obtained in the form of overseas ground terminals contributed by those countries. International exploitation and enhancement of the meteorological experiments through the synchronized participation of some 35 foreign nations represent another by-product of the applications program and one of particular interest to the less developed nations, including the neutrals, and even certain of the Soviet bloc satellite nations.

These international activities do not in most cases require special funding; indeed, they have brought participation resulting in modest savings. Nevertheless, this program of technical and political value can be maintained only as an extension of the underlying on-going programs, many of which are not considered part of the manned lunar landing program, but of importance to space science and direct applications.

Summary and Conclusion

In summarizing the views which are held by Dr. Dryden, Dr. Seamans, and myself, and which have guided our joint efforts to develop the National Space Program, I would emphasize that the manned lunar landing program, although of highest national priority, will not by itself create the pre-eminent position we seek. The present interest of the United States in terms of our scientific posture and increasing prestige, and our future interest in terms of having an adequate scientific and technological base for space activities beyond the manned lunar landing, demand that we pursue an adequate, well-balanced space program in all areas, including those not directly related to the manned lunar landing. We strongly believe that the United States will gain tangible benefits from such a total accumulation of basic scientific and technological data as well as from the greatly increased strength of our educational institutions. For these reasons, we believe it would not be in the nation's long-range interest to cancel or drastically curtail on-going space science and technology development programs in order to increase the funding of the manned lunar landing program in fiscal year 1963.

The fiscal year 1963 budget for major hardware development and flight missions not part of the manned lunar landing program, as well as the university program, totals $400 million. This is the amount which the manned space flight program is short. Cancellation of this effort would eliminate all nuclear developments, our international sounding rocket projects, the joint U.S.-Italian San Marcos project recently signed by Vice

President Johnson, all of our planetary and astronomical flights, and the communication and meteorological satellites. It should be realized that savings to the Government from this cancellation would be a small fraction of this total since considerable effort has already been expended in fiscal year 1963. However, even if the full amount could be realized, we would strongly recommend against this action.

In aeronautical and space research, we now have a program under way that will insure that we are covering the essential areas of the "unknown." Perhaps of one thing only can we be certain; that the ability to go into space and return at will increases the likelihood of new basic knowledge on the order of the theory that led to nuclear fission.

Finally, we believe that a supplemental appropriation for fiscal year 1963 is not nearly so important as to obtain for fiscal year 1964 the funds needed for the continued vigorous prosecution of the manned lunar landing program ($4.6 billion) *and* for the continuing development of our program in space science ($670 million), advanced research and technology ($263 million), space application ($185 million), and advanced manned flight including nuclear propulsion ($485 million). The funds already appropriated permit us to maintain a driving, vigorous program in the manned space flight area aimed at a target date of late 1967 for the lunar landing. We are concerned that the efforts required to pass a supplemental bill through the Congress, coupled with Congressional reaction to the practice of deficiency spending, could adversely affect our appropriations for fiscal year 1964 and subsequent years, and permit critics to focus on such items as charges that "overruns stem from poor management" instead of on the tremendous progress we have made and are making.

As you know, we have supplied the Bureau of the Budget complete information on the work that can be accomplished at various budgetary levels running from $5.2 billion to $6.6 billion for fiscal year 1964. We have also supplied the Bureau of the Budget with carefully worked out schedules showing that approval by you and the Congress of a 1964 level of funding of $6.2 billion together with careful husbanding and management of the $3.7 billion appropriated for 1963 would permit maintenance of the target dates necessary for the various milestones required for a final target date for the lunar landing of late 1967. The jump from $3.7 billion for 1963 to $6.2 billion for 1964 is undoubtedly going to raise more questions than the previous year jump from $1.8 billion to $3.7 billion.

If your budget for 1964 supports our request for $6.2 billion for NASA, we feel reasonably confident we can work with the committees and leaders of Congress in such a way as to secure their endorsement of your recommendation and the incident appropriations. To have moved in two years from President Eisenhower's appropriation request for 1962 of $1.1 billion to the approval of your own request for $1.8 billion, then for $3.7 billion for 1963 and on to $6.2 billion for 1964 would represent a great accomplishment for your administration. We see a risk that this will be lost sight of in charges that the costs are skyrocketing, the program is not under control, and so forth, if we request a supplemental in fiscal year 1963.

However, if it is your feeling that additional funds should be provided through a supplemental appropriation request for 1963 rather than to make the main fight for the level of support of the program on the basis of the $6.2 billion request for 1964, we will give our best effort to an effective presentation and effective use of any funds provided to speed up the manned lunar program.

With much respect, believe me

Sincerely yours,

James E. Webb
Administrator

Appendix 3

Summary of My Eyes-Only Draft Memorandum to Mr. Webb, 15 December 1966

The subject of this memo was NASA management. The first paragraph states, "In considering your questions relating to my views on organization as contained in your draft memo of 11/16/66, I found it helpful to start thinking of individual relationships, then to analyze organizational structure and communications in general, before reviewing specific changes that might improve NASA's effectiveness." The 14-page discourse rationalized the need for a new Associate Administrator for Management and Administration. Ultimately, the function was named the Office for Organization and Management. How these ideas were combined with those of Jim Webb and Harry Finger is discussed on pages 73, 91, and 92 of the text. The responsibilities of the proposed Associate Administrator are summarized below in an attachment to the 15 December memorandum.

Management and Administration

The Administrator, Deputy Administrator, Associate Deputy Administrator, the four Associate Administrators, and the Assistant Administrators will look to this office for support, advice, evaluation, and direct action where specified in matters related to the internal administration of the agency. The functions to be grouped in this office are indicated below. Seven main groups are now envisaged within the new office, with the tentative names and functions listed below.

1. The Resource Administration
   a. *Budget*—perform all budget functions (see list below); support AA's in their functions with co-located staffs; and support the new administrator for Program Planning and Analysis in the Office of the Administrator on budget and related matters.

Formulation and review of budget estimates and operating plans

Preparation of BOB and Congressional submissions

Development and review of operating plans and 506's and relative documents

Review resource implications of Program Approval Documents

Monitoring and review of recurring financial and program reports

Conduct financial operations

Liaison and point of contact on budgeting and programming matters with BOB and Congressional committees

   b. *Systems*—develop and supervise all agency-wide management, information, and control systems (see list below); support AA's in their functions, with co-located staffs where appropriate; and support and be responsive to the needs of the new administrator for Program Planning and Analysis in the Office of the Administrator.

Programming

Budgeting

Accounting

Manpower

Documentation

Agency Reports

Other information and control systems

2. *Manpower and Personnel*—perform manpower and personnel functions (see list below); support AA's with co-located staffs if necessary and appropriate.

Manpower planning, review, allocations, and controls

Personnel planning and operations

Training

Health

Related functions

3. *Industry Affairs*—responsible for:

Procurement

Labor relations

Inventions and contributions

4. *Institutional Development and Support*—responsible for:

Facilities management

Transportation and logistics

Property and supply

Security

Safety

Occupational medicine

Technical support (reliability and quality assurance)

5. *Compliance and Appraisal*—responsible for:

Technical and general evaluations

Audit

Inspections

6. *Headquarters and Administration*

7. *Management Analysis*

About the Author

D r. Robert C. Seamans, Jr., was born on 30 October 1918 in Salem, Massachusetts. He earned a bachelor of science degree in engineering at Harvard University in 1939, a master of science degree in aeronautics at Massachusetts Institute of Technology (MIT) in 1942, and a doctor of science degree in instrumentation from MIT in 1951.

From 1941 to 1955, he held teaching and project positions at MIT, during which time he worked on aeronautical problems, including instrumentation and control of airplanes and missiles. Dr. Seamans joined RCA in 1955 as manager of the Airborne Systems Laboratory and chief systems engineer of the Airborne Systems Department. From 1948 to 1958, Dr. Seamans also served on technical committees of NASA's predecessor organization, the National Advisory Committee for Aeronautics.

In 1960, Dr. Seamans joined NASA as Associate Administrator, part of the top management "Triad" with Hugh Dryden and James Webb. In 1965, he became Deputy Administrator, retaining many of the general management-type responsibilities of the Associate Administrator and also serving as Acting Administrator. In January 1968, he resigned from NASA to become a visiting professor at MIT.

He was named Secretary of the United States Air Force in 1969, serving until 1973. Dr. Seamans was also president of the National Academy of Engineering from May 1973 to December 1974, when he became the first administrator of the new Energy Research and Development Administration. He returned to MIT in 1977, becoming dean of its School of Engineering in 1978. He has served on many prestigious blue ribbon commissions since then.

Dr. Seamans and his wife, Eugenia A. "Gene" Merrill, have 5 children and 12 grandchildren.

To learn more about Robert C. Seamans, Jr., see his autobiography, *Aiming at Targets* (NASA SP-4106, 1996).

Acronyms and Abbreviations

AACB	Aeronautics and Astronautics Coordinating Board
AAS	American Astronautical Society
AEC	Atomic Energy Commission
AGARD	Advisory Group for Aerospace Research & Development
AIAA	American Institute of Aeronautics and Astronautics
AMR	Atlantic Mission Range
ATDA	Augmented Target Docking Adapter
BMEWS	Ballistic Missile Early Warning System
BoB	Bureau of the Budget (now the Office of Management and Budget)
C-1, C-2	configurations for the Nova launch vehicle
C-2, C-3	Saturn configurations
Caltech	California Institute of Technology
CEV	Crew Exploration Vehicle
CIA	Central Intelligence Agency
DOD	Department of Defense
DX	label for a high-priority program
EST	eastern standard time
EVA	extravehicular activity
FCRC	Federal Contracted Research Center
FFRDC	Federally Funded Research and Development Center
fps	feet per second
FY	fiscal year
GAO	Government Accounting Office
GE	General Electric

HSS-2	type of helicopter
IBEW	International Brotherhood of Electrical Workers
IBM	International Business Machines
ICBM	intercontinental ballistic missile
ICD	Interface Control Documents
ISS	International Space Station
JPL	Jet Propulsion Laboratory
LEM	lunar excursion module
LOR	lunar orbit rendezvous
MA	Mercury capsule with Atlas booster
max-q	maximum dynamic pressure
MIT	Massachusetts Institute of Technology
mph	miles per hour
MQF	Mobile Quarantine Facility
MR	Mercury Redstone rocket (usually with number, e.g., MR-3)
MSC	Manned Spacecraft Center
MSFC	Marshall Space Flight Center
NAA	North American Aviation
NACA	National Advisory Committee for Aeronautics
NASA	National Aeronautics and Space Administration
NATO	North Atlantic Treaty Organisation
NBC	National Broadcasting Company
OMB	Office of Management and Budget
OMSF	Office of Manned Spaceflight

PAD	Project Approval Document
POGO effect	longitudinal vibrations (as in the movement of a pogo stick)
psi	pounds per square inch
R&D	research and development
R&R	rest and relaxation
RCA	Radio Corporation of America
RFP	request for proposal
RFQ	request for quotation
RIF	reduction in force
SAINT	SAtellite INTerceptor
SEPC	Space Exploration Program Council
SMS	Sequenced Milestone System
TIROS	Television Infrared Observation Satellite Program
TRW	Thompson Ramo Woldridge
UN	United Nations
USSR	Union of Soviet Socialist Republics
V-1, V-2	German "vengeance" weapons
VAB	Vertical Assembly Building

NASA Monographs in Aerospace History Series

All monographs except the first one are available by sending a self-addressed 9-by-12-inch envelope for each monograph with appropriate postage for 15 ounces to the NASA History Division, Room CO72, Washington, DC 20546. A complete listing of all NASA History Series publications is available at *http://history.nasa.gov/series95.html* on the World Wide Web. In addition, a number of monographs and other History Series publications are available online from the same URL.

Launius, Roger D., and Aaron K. Gillette, compilers. *Toward a History of the Space Shuttle: An Annotated Bibliography.* Monographs in Aerospace History, No. 1, 1992. Out of print.

Launius, Roger D., and J. D. Hunley, compilers. *An Annotated Bibliography of the Apollo Program.* Monographs in Aerospace History, No. 2, 1994.

Launius, Roger D. Apollo: *A Retrospective Analysis.* Monographs in Aerospace History, No. 3, 1994.

Hansen, James R. *Enchanted Rendezvous: John C. Houbolt and the Genesis of the Lunar-Orbit Rendezvous Concept.* Monographs in Aerospace History, No. 4, 1995.

Gorn, Michael H. *Hugh L. Dryden's Career in Aviation and Space.* Monographs in Aerospace History, No. 5, 1996.

Powers, Sheryll Goecke. *Women in Flight Research at NASA Dryden Flight Research Center from 1946 to 1995.* Monographs in Aerospace History, No. 6, 1997.

Portree, David S. F., and Robert C. Trevino. *Walking to Olympus: An EVA Chronology.* Monographs in Aerospace History, No. 7, 1997.

Logsdon, John M., moderator. *Legislative Origins of the National Aeronautics and Space Act of 1958: Proceedings of an Oral History Workshop.* Monographs in Aerospace History, No. 8, 1998.

Rumerman, Judy A., compiler. *U.S. Human Spaceflight, A Record of Achievement 1961–1998.* Monographs in Aerospace History, No. 9, 1998.

Portree, David S. F. *NASA's Origins and the Dawn of the Space Age.* Monographs in Aerospace History, No. 10, 1998.

Logsdon, John M. *Together in Orbit: The Origins of International Cooperation in the Space Station.* Monographs in Aerospace History, No. 11, 1998.

Phillips, W. Hewitt. *Journey in Aeronautical Research: A Career at NASA Langley Research Center.* Monographs in Aerospace History, No. 12, 1998.

Braslow, Albert L. *A History of Suction-Type Laminar-Flow Control with Emphasis on Flight Research.* Monographs in Aerospace History, No. 13, 1999.

Logsdon, John M., moderator. *Managing the Moon Program: Lessons Learned From Apollo.* Monographs in Aerospace History, No. 14, 1999.

Perminov, V. G. *The Difficult Road to Mars: A Brief History of Mars Exploration in the Soviet Union.* Monographs in Aerospace History, No. 15, 1999.

Tucker, Tom. *Touchdown: The Development of Propulsion Controlled Aircraft at NASA Dryden.* Monographs in Aerospace History, No. 16, 1999.

Maisel, Martin, Demo J. Giulanetti, and Daniel C. Dugan, *The History of the XV-15 Tilt Rotor Research Aircraft: From Concept to Flight.* Monographs in Aerospace History, No. 17, 2000 (NASA SP-2000-4517).

Jenkins, Dennis R. *Hypersonics Before the Shuttle: A Concise History of the X-15 Research Airplane.* Monographs in Aerospace History, No. 18, 2000 (NASA SP-2000-4518).

Chambers, Joseph R. *Partners in Freedom: Contributions of the Langley Research Center to U.S. Military Aircraft of the 1990s.* Monographs in Aerospace History, No. 19, 2000 (NASA SP-2000-4519).

Waltman, Gene L. *Black Magic and Gremlins: Analog Flight Simulations at NASA's Flight Research Center.* Monographs in Aerospace History, No. 20, 2000 (NASA SP-2000-4520).

Portree, David S. F. *Humans to Mars: Fifty Years of Mission Planning, 1950–2000.* Monographs in Aerospace History, No. 21, 2001 (NASA SP-2001-4521).

Thompson, Milton O., with J. D. Hunley. *Flight Research: Problems Encountered and What They Should Teach Us.* Monographs in Aerospace History, No. 22, 2001 (NASA SP-2001-4522).

Tucker, Tom. *The Eclipse Project.* Monographs in Aerospace History, No. 23, 2001 (NASA SP-2001-4523).

Siddiqi, Asif A. *Deep Space Chronicle: A Chronology of Deep Space and Planetary Probes, 1958–2000.* Monographs in Aerospace History, No. 24, 2002 (NASA SP-2002-4524).

Merlin, Peter W. *Mach 3+: NASA/USAF YF-12 Flight Research, 1969–1979.* Monographs in Aerospace History, No. 25, 2001 (NASA SP-2001-4525).

Anderson, Seth B. *Memoirs of an Aeronautical Engineer: Flight Tests at Ames Research Center: 1940–1970.* Monographs in Aerospace History, No. 26, 2002 (NASA SP-2002-4526).

Renstrom, Arthur G. *Wilbur and Orville Wright: A Bibliography Commemorating the One-Hundredth Anniversary of the First Powered Flight on December 17, 1903.* Monographs in Aerospace History, No. 27, 2002 (NASA SP-2002-4527).

Chambers, Joseph R. *Concept to Reality: Contributions of the NASA Langley Research Center to U.S. Civil Aircraft of the 1990s.* Monographs in Aerospace History, No. 29, 2003 (NASA SP-2003-4529).

Peebles, Curtis, editor. *The Spoken Word: Recollections of Dryden History, The Early Years.* Monographs in Aerospace History, No. 30, 2003 (NASA SP-2003-4530).

Jenkins, Dennis R., Tony Landis, and Jay Miller. *American X-Vehicles: An Inventory—X-1 to X-50.* Monographs in Aerospace History, No. 31, 2003 (NASA SP-2003-4531).

Renstrom, Arthur G. *Wilbur and Orville Wright Chronology.* Monographs in Aerospace History, No. 32, 2003 (NASA SP-2003-4532).

Bowles, Mark D., and Robert S. Arrighi. *NASA's Nuclear Frontier: The Plum Brook Reactor Facility, 1941–2002.* Monographs in Aerospace History, No. 33, 2004 (NASA SP-2004-4533).

McCurdy, Howard E. *Low-Cost Innovation in Spaceflight: The Near Earth Asteroid Rendezvous (NEAR) Shoemaker Mission.* Monographs in Aerospace History, No. 36, 2005 (NASA SP-2005-4536).

Index